Lecture Notes in Mathematics

For information about Vols. 1–521, please contact your bookseller or Springer-Verlag.

Vol. 522: C. O. Bloom and N. D. Kazarinoff, Short Wave Radiation Problems in Inhomogeneous Media: Asymptotic Solutions. V. 104 pages. 1976.

Vol. 523: S. A. Albeverio and R. J. Høegh-Krohn, Mathematical Theory of Feynman Path Integrals. IV, 139 pages. 1976.

Vol. 524: Séminaire Pierre Lelong (Analyse) Année 1974/75. Edité par P. Lelong. V, 222 pages. 1976.

Vol. 525: Structural Stability, the Theory of Catastrophes, and Applications in the Sciences. Proceedings 1975. Edited by P. Hilton. VI, 408 pages. 1976.

Vol. 526: Probability in Banach Spaces. Proceedings 1975. Edited by A. Beck. VI, 290 pages. 1976.

Vol. 527: M. Denker, Ch. Grillenberger, and K. Sigmund, Ergodic Theory on Compact Spaces. IV, 360 pages. 1976.

Vol. 528: J. E. Humphreys, Ordinary and Modular Representations of Chevalley Groups. III, 127 pages. 1976.

Vol. 529: J. Grandell, Doubly Stochastic Poisson Processes. X, 234 pages. 1976.

Vol. 530: S. S. Gelbart, Weil's Representation and the Spectrum of the Metaplectic Group. VII, 140 pages. 1976.

Vol. 531: Y.-C. Wong, The Topology of Uniform Convergence on Order-Bounded Sets. VI, 163 pages. 1976.

Vol. 532: Théorie Ergodique. Proceedings 1973/1974. Edité par J.-P. Conze and M. S. Keane. VIII, 227 pages. 1976.

Vol. 533: F. R. Cohen, T. J. Lada, and J. P. May, The Homology of Iterated Loop Spaces. IX, 490 pages. 1976.

Vol. 534: C. Preston, Random Fields. V, 200 pages. 1976.

Vol. 535: Singularités d'Applications Differentiables. Plans-sur-Bex. 1975. Edité par O. Burlet et F. Ronga. V, 253 pages. 1976.

Vol. 536: W. M. Schmidt, Equations over Finite Fields. An Elementary Approach. IX, 267 pages. 1976.

Vol. 537: Set Theory and Hierarchy Theory. Bierutowice, Poland 1975. A Memorial Tribute to Andrzej Mostowski. Edited by W. Marek, M. Srebrny and A. Zarach. XIII, 345 pages. 1976.

Vol. 538: G. Fischer, Complex Analytic Geometry. VII, 201 pages. 1976.

Vol. 539: A. Badrikian, J. F. C. Kingman et J. Kuelbs, Ecole d'Eté de Probabilités de Saint Flour V-1975. Edité par P.-L. Hennequin. IX, 314 pages. 1976.

Vol. 540: Categorical Topology, Proceedings 1975. Edited by E. Binz and H. Herrlich. XV, 719 pages. 1976.

Vol. 541: Measure Theory, Oberwolfach 1975. Proceedings. Edited by A. Bellow and D. Kölzow. XIV, 430 pages. 1976.

Vol. 542: D. A. Edwards and H. M. Hastings, Čech and Steenrod Homotopy Theories with Applications to Geometric Topology. VII, 296 pages. 1976.

Vol. 543: Nonlinear Operators and the Calculus of Variations, Bruxelles 1975. Edited by J. P. Gossez, E. J. Lami Dozo, J. Mawhin, and L. Waelbroeck, VII, 237 pages. 1976.

Vol. 544: Robert P. Langlands, On the Functional Equations Satisfied by Eisenstein Series. VII, 337 pages. 1976.

Vol. 545: Noncommutative Ring Theory. Kent State 1975. Edited by J. H. Cozzens and F. L. Sandomierski. V, 212 pages. 1976.

Vol. 546: K. Mahler, Lectures on Transcendental Numbers. Edited and Completed by B. Diviš and W. J. Le Veque. XXI, 254 pages. 1976.

Vol. 547: A. Mukherjea and N. A. Tserpes, Measures on Topological Semigroups: Convolution Products and Random Walks. V, 197 pages. 1976.

Vol. 548: D. A. Hejhal, The Selberg Trace Formula for PSL $(2,\mathbb{R})$. Volume I. VI, 516 pages. 1976.

Vol. 549: Brauer Groups, Evanston 1975. Proceedings. Edited by D. Zelinsky. V, 187 pages. 1976.

Vol. 550: Proceedings of the Third Japan – USSR Symposium on Probability Theory. Edited by G. Maruyama and J. V. Prokhorov. VI, 722 pages. 1976.

Vol. 551: Algebraic K-Theory, Evanston 1976. Proceedings. Edited by M. R. Stein. XI, 409 pages. 1976.

Vol. 552: C. G. Gibson, K. Wirthmüller, A. A. du Plessis and E. J. N. Looijenga. Topological Stability of Smooth Mappings. V, 155 pages. 1976.

Vol. 553: M. Petrich, Categories of Algebraic Systems. Vector and Projective Spaces, Semigroups, Rings and Lattices. VIII, 217 pages. 1976.

Vol. 554: J. D. H. Smith, Mal'cev Varieties. VIII, 158 pages. 1976.

Vol. 555: M. Ishida, The Genus Fields of Algebraic Number Fields. VII, 116 pages. 1976.

Vol. 556: Approximation Theory. Bonn 1976. Proceedings. Edited by R. Schaback and K. Scherer. VII, 466 pages. 1976.

Vol. 557: W. Iberkleid and T. Petrie, Smooth S^1 Manifolds. III, 163 pages. 1976.

Vol. 558: B. Weisfeiler, On Construction and Identification of Graphs. XIV, 237 pages. 1976.

Vol. 559: J.-P. Caubet, Le Mouvement Brownien Relativiste. IX, 212 pages. 1976.

Vol. 560: Combinatorial Mathematics, IV, Proceedings 1975. Edited by L. R. A. Casse and W. D. Wallis. VII, 249 pages. 1976.

Vol. 561: Function Theoretic Methods for Partial Differential Equations. Darmstadt 1976. Proceedings. Edited by V. E. Meister, N. Weck and W. L. Wendland. XVIII, 520 pages. 1976.

Vol. 562: R. W. Goodman, Nilpotent Lie Groups: Structure and Applications to Analysis. X, 210 pages. 1976.

Vol. 563: Séminaire de Théorie du Potentiel. Paris, No. 2. Proceedings 1975–1976. Edited by F. Hirsch and G. Mokobodzki. VI, 292 pages. 1976.

Vol. 564: Ordinary and Partial Differential Equations, Dundee 1976. Proceedings. Edited by W. N. Everitt and B. D. Sleeman. XVIII, 551 pages. 1976.

Vol. 565: Turbulence and Navier Stokes Equations. Proceedings 1975. Edited by R. Temam. IX, 194 pages. 1976.

Vol. 566: Empirical Distributions and Processes. Oberwolfach 1976. Proceedings. Edited by P. Gaenssler and P. Révész. VII, 146 pages. 1976.

Vol. 567: Séminaire Bourbaki vol. 1975/76. Exposés 471–488. IV, 303 pages. 1977.

Vol. 568: R. E. Gaines and J. L. Mawhin, Coincidence Degree, and Nonlinear Differential Equations. V, 262 pages. 1977.

Vol. 569: Cohomologie Etale SGA 4½. Séminaire de Géométrie Algébrique du Bois-Marie. Edité par P. Deligne. V, 312 pages. 1977.

Vol. 570: Differential Geometrical Methods in Mathematical Physics, Bonn 1975. Proceedings. Edited by K. Bleuler and A. Reetz. VIII, 576 pages. 1977.

Vol. 571: Constructive Theory of Functions of Several Variables, Oberwolfach 1976. Proceedings. Edited by W. Schempp and K. Zeller. VI. 290 pages. 1977

Vol. 572: Sparse Matrix Techniques, Copenhagen 1976. Edited by V. A. Barker. V, 184 pages. 1977.

Vol. 573: Group Theory, Canberra 1975. Proceedings. Edited by R. A. Bryce, J. Cossey and M. F. Newman. VII, 146 pages. 1977.

Vol. 574: J. Moldestad, Computations in Higher Types. IV, 203 pages. 1977.

Vol. 575: K-Theory and Operator Algebras, Athens, Georgia 1975. Edited by B. B. Morrel and I. M. Singer. VI, 191 pages. 1977.

Vol. 576: V. S. Varadarajan, Harmonic Analysis on Real Reductive Groups. VI, 521 pages. 1977.

Vol. 577: J. P. May, E_∞ Ring Spaces and E_∞ Ring Spectra. IV, 268 pages. 1977.

Vol. 578: Séminaire Pierre Lelong (Analyse) Année 1975/76. Edité par P. Lelong. VI, 327 pages. 1977.

Vol. 579: Combinatoire et Représentation du Groupe Symétrique, Strasbourg 1976. Proceedings 1976. Edité par D. Foata. IV, 339 pages. 1977.

continuation on page 141

Lecture Notes in Mathematics

Edited by A. Dold and B. Eckmann

731

Yoshiomi Nakagami
Masamichi Takesaki

Duality for Crossed Products of von Neumann Algebras

Springer-Verlag
Berlin Heidelberg New York 1979

Authors

Yoshiomi Nakagami
Department of Mathematics
Yokohama City University
Yokohama
Japan

Masamichi Takesaki
Department of Mathematics
University of California
Los Angeles, CA 90024
U.S.A.

AMS Subject Classifications (1970): 46L10

ISBN 3-540-09522-5 Springer-Verlag Berlin Heidelberg New York
ISBN 0-387-09522-5 Springer-Verlag New York Heidelberg Berlin

Library of Congress Cataloging in Publication Data
Nakagami, Yoshiomi, 1940-
Duality for crossed products of von Neumann algebras.
(Lecture notes in mathematics ; 731)
Bibliography: p.
Includes index.
1. Von Neumann algebras--Crossed products.
2. Duality theory (Mathematics) I. Takesaki, Masamichi, 1933- II. Title. III. Series:
Lecture notes in Mathematics (Berlin) 731.
QA3.L28 no. 731 [QA326] 510'.8s [512'.55] 79-17038
ISBN 0-387-09522-5

Printing and binding: Beltz Offsetdruck, Hemsbach/Bergstr.
2141/3140-543210

INTRODUCTION

The recent development in the theory of operator algebras showed the importance of the study of automorphism groups of von Neumann algebras and their crossed products. The main tool here is duality theory for locally compact groups.

Let $\mathfrak{M}$ be a von Neumann algebra equipped with a continuous action α of a locally compact group G. For a unitary representation $\{U,\mathfrak{H}_U\}$ of G, let $\mathfrak{M}^{\alpha}(U)$ be the σ-weakly closed subspace of $\mathfrak{M}$ spanned by the range of all intertwining operators T from $\mathfrak{H}_U$ into $\mathfrak{M}$. It is easily seen that $\mathfrak{M}^{\alpha}(U)\mathfrak{M}^{\alpha}(V)$ is contained in $\mathfrak{M}^{\alpha}(U \otimes V)$ for any pair U,V of unitary representations of G, and that $\mathfrak{M}^{\alpha}(U)^* = \mathfrak{M}^{\alpha}(\overline{U})$ where $\overline{U}$ means the conjugate representation of U. This simple fact is the basis for the entire duality mechanism. At this point, one should recall the formulation of the Tannaka-Tatsuuma duality theorem.

In spite of the above simple basis, the absence of the dual group in the non-commutative case forces us to employ the notationally (if not mathematically) complicated Hopf-von Neumann algebra approach to the duality principle. It should however be pointed out that the Hopf - von Neumann algebra approach simply means a systematic usage of the unitary W_G on $L^2(G \times G)$ given by $(W_G\xi)(s,t) = \xi(s,ts)$. This operator W_G is nothing else but the operator version of the group multiplication table. In this sense, W_G is a very natural object whose importance can not be overestimated. For example, the Tannaka-Tatsuuma duality theorem simply asserts that a non-zero $x \in \mathcal{L}(L^2(G))$ is of the form $x = \rho(t)$, where ρ is the right regular representation, if and only if $W_G^*(x \otimes 1)W_G = x \otimes x$.

When the crossed product of an operator algebra was introduced by Turumaru, [76], Suzuki, [61], Nakamura-Takeda, [51,52], Doplicher-Kastler-Robinson, [20] and Zeller-Meier, [79], it was considered as a method to construct a new algebra from a given covariant system, although Doplicher-Kastler-Robinson's work was directed more toward the construction of covariant representations. Thus it was hoped to add more new examples as it was the case for Murray and von Neumann in the group measure space construction. In the course of the structure analysis of factors of type III, it was recognized [12] that the study of crossed products is indeed the study of a special class of perturbations of an action α on $\mathfrak{M}$ by means of integrable 1-cocycles. More precisely, the crossed product $\mathfrak{M} \times_{\alpha} G$ is precisely the fixed point algebra $\widetilde{\mathfrak{M}}^{\tilde{\alpha}}$ in the von Neumann algebra $\overline{\mathfrak{M}} = \mathfrak{M} \bar{\otimes} \mathcal{L}(L^2(G))$ under the new action $\tilde{\alpha}_s = \alpha_s \otimes \mathrm{Ad}(\lambda(s))$, where λ is the left regular representation. With this observation, Connes and Takesaki viewed the theory of crossed products as the study of the perturbed action by the regular representation, [14]; thus they proposed the comparison theory of 1-cocycles as a special application of the Murray - von Neumann dimension theory for von Neumann algebras.

In this setting, the duality principle for non-commutative groups comes into play in a natural fashion as pointed out above. Suppose at the moment that G is abelian. If α is a "good" action of G on $\mathfrak{M}$, so that for each $p \in \hat{G}$ one can choose a unitary u such that $\alpha_s(u) = \langle s,p \rangle u$, then this unitary u gives rise to an action of p on the fixed point algebra $\mathfrak{M}^\alpha$; or these u's together with $\mathfrak{M}^\alpha$ generate $\mathfrak{M}$. If we drop the commutativity assumption from G, then $\hat{G}$ should be replaced by something else. There are a few candidates. One is the algebra $L^\infty(G)$ together with the co-multiplication α_G given by $\alpha_G(f)(s,t) = f(st)$; the second is the Fourier algebra A(G) , the predual of the von Neumann algebra $\mathcal{R}(G)$ generated by $\rho(G)$, [27,28]; the third is the ring of unitary representations. At any rate, it will be shown that if the action α is "good", then the "dual" of G acts on $\mathfrak{M}^\alpha$ and $\mathfrak{M}$ is generated by the "dual" of G and $\mathfrak{M}^\alpha$. The precise meaning of an action of the "dual" will be given as a co-action δ of G as well as a Roberts action of a ring of representations. The crossed product of a von Neumann algebra $\mathfrak{n}$ by an action of the "dual", a co-action and a Roberts action, is formulated in Chapter IV and duality theorems, Theorems I.2.5 and I.2.7, are proved there. The equivalence of co-actions and Roberts actions is established in §4 in Chapter IV.

In this paper, we present the dualized version of the Arveson-Connes spectral analysis, the integrability of an action, dominant actions and the comparison theory of 1-cocycles. As an application of our theory, the Galois type correspondence between closed normal subgroups of G and certain von Neumann algebras containing $\mathfrak{M}^\alpha$ is established in Chapter VII. We must point out that the restriction of the normality for subgroups should be lifted through an application of Fell's theory of Banach *-algebra bundles, [30], and its dualized version. We shall treat this somewhere else.

The present notes have grown out of an attempt to give an expository unified account of the present stage of the theory of crossed products for the International Conference on C*-algebras and their Applications to Theoretical Physics, CNRS, Marseille, June 1977. In theoretical physics, the analysis of the fixed point algebra is particularly relevant to the theory of gauge groups and/or the reconstruction of the field algebra out of the observable algebra. In this respect, the material presented in Chapter VII as well as those related to Roberts actions are relevant for the reader motivated by physics. It should, however, be mentioned that a theory concerning C*-algebras is more needed in theoretical physics. It is indeed a very active area. The authors hope that the present notes will set a platform for the further development.

The present notes are written in expository style, while Chapters III, IV and V are partially new. The references are cited at the end of each section.

The authors would like to express their sincere gratitude to Prof. D. Kastler and his colleagues at CNRS, Marseille, for their warm hospitality extended to them while this work was prepared.

CONTENTS

LIST OF SYMBOLS

$\mathbb{N}$ = The set of natural numbers, $\{1,2,\dots\}$.

$\mathbb{Z}$ = The ring of integers.

$\mathbb{Q}$ = The rational number field.

$\mathbb{R}$ = The real number field.

$\mathbb{C}$ = The complex number field.

$\mathbb{Z}_+, \mathbb{R}_+$: The non-negative parts.

$\mathfrak{H}, \mathfrak{K}, \dots$: Hilbert spaces.

$\mathfrak{M}, \mathfrak{N}, \dots$: Subspaces.

$\xi, \eta, \zeta, \dots$: Vectors in a Hilbert space.

$\varepsilon_1, \varepsilon_2, \dots$: Vectors in a fixed orthogonal normalized basis.

G : A locally compact group.

$L^2(G)$ = The Hilbert space of all square integrable functions with respect to a right invariant Haar measure ds on G.

$L^\infty(G)$ = The abelian von Neumann algebra of all essentially bounded functions on G with respect to the Haar measure acting on $L^2(G)$ by multiplication.

$\mathcal{L}(\mathfrak{H}), \mathcal{L}(\mathfrak{K}), \dots$: The algebra of all bounded operators.

$\mathcal{M}, \mathcal{N}, \mathcal{P}, \mathcal{Q}, \dots$: von Neumann algebras. Unless otherwise stated, $\mathcal{M}$ acts on $\mathfrak{H}$, and $\mathcal{N}$ acts on $\mathfrak{K}$.

$\mathcal{A}, \mathcal{B}, \dots$: Abelian von Neumann algebras.

$\mathcal{C}_\mathcal{M}, \mathcal{C}_\mathcal{N}, \dots$: The center of $\mathcal{M}$ and $\mathcal{N}$ respectively. $\mathcal{M} \vee \mathcal{N} = (\mathcal{M} \cup \mathcal{N})''$.

$e, f, \dots, p, q, \dots$: Projections.

$\mathrm{Aut}(\mathcal{M})$ = The group of automorphisms (*-preserving) of $\mathcal{M}$.

$\mathrm{Aut}(\mathcal{M}/\mathcal{N})$ = The group of automorphisms of $\mathcal{M}$ leaving the von Neumann subalgebra $\mathcal{N}$ of $\mathcal{M}$ pointwise fixed.

ι = The identity automorphism.

σ = The symmetry reflection: $x \otimes y \to y \otimes x$.

τ, Tr = Traces.

$\omega, \varphi, \psi, \dots$: Linear functionals, states or weights.

$\varphi x, y\varphi$: $\langle xy, \varphi\rangle = \langle y, \varphi x\rangle = \langle x, y\varphi\rangle$.

$e \in G$: The unit.

$r, s, t, \dots$: Elements in G.

$\rho(\cdot)$ = The right regular representation of G.

$\lambda(\cdot)$ = The left regular representation of G.

$\rho_t(x) = \rho(t)\, x\, \rho(t)^*$ for each $x \in \mathcal{L}(L^2(G))$, $t \in G$.

$\lambda_t(x) = \lambda(t)\, x\, \lambda(t)^*$ for each $x \in \mathcal{L}(L^2(G))$, $t \in G$.

$$f * g(t) = \int_G f(ts^{-1}) g(s)\, ds \ ;$$

$$f^\#(t) = \overline{f(t^{-1})} \ ;$$

$$f^{\flat}(t) = \Delta(t)\overline{f(t^{-1})} \; ;$$

$$f^{\vee}(t) = f(t^{-1}) \; , \quad f^{\wedge}(t) = \Delta(t)f(t^{-1}) \; .$$

$\mathcal{R}(G) = \{\rho(t) : t \in G\}''$.

$\mathcal{R}'(G) = \{\lambda(t) : t \in G\}'' = \mathcal{R}(G)'$.

$\Delta(\cdot)$ = The modular function of G.

$A(G)$: The Fourier algebra of G, which is identified with $\mathcal{R}(G)_*$ by $\omega_{f,g}(\rho(t)) = (g^{\#} * f)(t)$, $t \in G$, i.e. $A(G) = L^2(G)^{\#} * L^2(G)$.

α,β,γ ... : Actions of G on a von Neumann algebra: $(\alpha\otimes\iota)\circ\alpha = (\iota\otimes\alpha_G)\circ\alpha$.

δ,ε, ... : Co-actions of G on a von Neumann algebra: $(\delta\otimes\iota)\circ\delta = (\iota\otimes\delta_G)\circ\delta$.

α_G : The action of G on $L^\infty(G)$ with $(\alpha_G)_t = \rho_t$; $(\alpha_G f)(s,t) = f(st)$.

α'_G : The isomorphism of $L^\infty(G)$ into $L^\infty(G)\bar{\otimes}L^\infty(G)$ with $(\alpha'_G)_t = \lambda_t^{-1}$; $(\alpha'_G f)(s,t) = f(ts)$.

δ_G : The co-action of G on $\mathcal{R}(G)$ with $\delta_G(\rho(t)) = \rho(t)\otimes\rho(t)$.

δ'_G : The co-action of G on $\mathcal{R}(G)'$ with respect to $\mathcal{R}(G)'$ such that $\delta'_G(\lambda(t)) = \lambda(t)\otimes\lambda(t)$, cf. Chapter I, § 4.

V_G, W_G, V'_G, W'_G : The unitaries on $L^2(G\times G) = L^2(G)\otimes L^2(G)$ defined by:

$$\begin{cases} V_G\xi(s,t) = \xi(st,t), W_G\xi(s,t) = \xi(s,ts) \; ; \\ V'_G\xi(s,t) = \Delta(t)^{\frac{1}{2}}\xi(t^{-1}s,t), W'_G\xi(s,t) = \Delta(s)^{\frac{1}{2}}\xi(s,s^{-1}t), \xi \in L^2(G\times G) \; . \end{cases}$$

α',β',... : The actions of G on a von Neumann algebra with respect to $\mathcal{R}'(G)$: $(\alpha'\otimes\iota)\circ\alpha' = (\iota\otimes\alpha'_G)\circ\alpha'$.

δ',ε',... : The co-actions of G on a von Neumann algebra with respect to $\mathcal{R}'(G)$: $(\delta'\otimes\iota)\circ\delta' = (\iota\otimes\delta'_G)\circ\delta'$.

$\mathfrak{M}^\alpha = \{x \in \mathfrak{M} : \alpha(x) = x\otimes 1\}$; $\mathfrak{N}^\delta = \{y \in \mathfrak{N} : \delta(y) = y\otimes 1\}$.

α^e: The action of G on $\mathfrak{M}_e$ with $\alpha^e(x_e) = \alpha(x)_{e\otimes 1}$ for $e \in \mathfrak{M}^\alpha$.

δ^e: The co-action of G on $\mathfrak{N}_e$ with $\delta^e(y_e) = \delta(y)_{e\otimes 1}$ for $e \in \mathfrak{N}^\delta$.

$\{\mathfrak{M},\alpha\}_e = \{\mathfrak{M}_e,\alpha^e\}$; $\{\mathfrak{N},\delta\}_e = \{\mathfrak{N}_e,\delta^e\}$.

$\mathfrak{n}_\varphi = \{x \in \mathfrak{M} : \varphi(x^*x) < \infty\}$; $\mathfrak{m}_\varphi = \mathfrak{n}_\varphi^*\mathfrak{n}_\varphi$.

$\{\pi_\varphi,\mathfrak{H}_\varphi\}$ or $\{\pi_\varphi,\mathfrak{H}_\varphi,\eta_\varphi\}$: The GNS construction: $(\pi_\varphi(x)\eta_\varphi(y)|\eta_\varphi(z)) = (xy|z)_\varphi = \varphi(z^*xy)$, $x \in \mathfrak{M}$, $y,z \in \mathfrak{n}_\varphi$.

π_r (resp. π_ℓ): The right (resp. left) representation of a right (resp. left) Hilbert algebra.

Δ_φ = The modular operator.

J_φ = The modular unitary involution.

σ^φ = The modular automorphism.

$\mathfrak{M}\times_\alpha G$ = The crossed product of $\mathfrak{M}$ by G with respect to α.

$\mathfrak{M}\times_{\alpha'} G$ = The crossed product of $\mathfrak{M}$ by G with respect to α'.

$\mathfrak{N}\times_\delta G$ = The crossed product of $\mathfrak{N}$ by G with respect to δ.

$\mathfrak{N}\times_{\delta'} G$ = The crossed product of $\mathfrak{N}$ by G with respect to δ'.

$\hat{\alpha}$ = The dual of α: $\hat{\alpha}(y) = \mathrm{Ad}_{1\otimes W_G^*}(y\otimes 1)$ for $y \in \mathfrak{M}\times_\alpha G$.

$\hat{\alpha}'$ = The dual of α': $\hat{\alpha}'(y) = \mathrm{Ad}_{1\otimes W'_G}(y \otimes 1)$ for $y \in \mathfrak{M} \times_{\alpha'} G$.

$\hat{\delta}$ = The dual of δ: $\hat{\delta}(x) = \mathrm{Ad}_{1\otimes V'_G}(x \otimes 1)$ for $x \in \mathfrak{N} \times_{\delta} G$.

$\hat{\delta}'$ = The dual of δ': $\hat{\delta}'(x) = \mathrm{Ad}_{1\otimes V^*_G}(x \otimes 1)$ for $x \in \mathfrak{N} \times_{\delta'} G$.

$\hat{\hat{\alpha}} = (\hat{\alpha})^\wedge$, $\hat{\hat{\delta}} = (\hat{\delta})^\wedge$.

$\bar{\alpha} = (\iota \otimes \sigma) \circ (\alpha \otimes \iota)$ and $\bar{\alpha}_t = \alpha_t \otimes \iota$.

$\bar{\delta} = (\iota \otimes \sigma) \circ (\delta \otimes \iota)$.

$\tilde{\alpha} = \mathrm{Ad}_{1\otimes V'_G} \circ \bar{\alpha}$ and $\tilde{\alpha}_t = \alpha_t \otimes \lambda_t$.

$\tilde{\delta} = \mathrm{Ad}_{1\otimes W_G} \circ \bar{\delta}$.

β = The action of G on $\alpha(\mathfrak{M})'$ with respect to $\mathcal{R}'(G)$: $\beta(x) = \mathrm{Ad}_{1\otimes V^*_G}(x \otimes 1)$.

ε = The co-action of G on $\delta(\mathfrak{N})'$: $\varepsilon(y) = \mathrm{Ad}_{1\otimes W_G}(y \otimes 1)$.

$C(G)$ = The set of continuous functions on G.

$C_\infty(G)$ = The set of continuous functions in $C(G)$ vanishing at ∞.

$\mathcal{K}(G)$ = The set of continuous functions in $C(G)$ vanishing outside a compact set.

$\mathrm{supp}(\varphi)$: The support of $\varphi \in A(G)$, which is the closure of the smallest set outside which φ vanishes.

μ_G, μ'_G: The weights on $L^\infty(G)$ defined by:

$$\mu_G(f) = \int f(t)dt \; ; \; \mu'_G(f) = \int f(t)d't ,$$

where $d't$ is the left invariant Haar measure $\Delta(t)dt$.

ψ_G, ψ'_G: The wights on $\mathcal{R}(G)$, $\mathcal{R}'(G)$ defined by:

$$\psi_G(\rho(f)) = f(e) \; ; \; \psi'_G(\lambda(f)) = f(e) \quad \text{for } A(G)_+ ,$$

where $A(G)_+$ is the set of positive definite functions in $A(G)$.

K,J: The operators on $L^2(G)$ defined by:

$$(Kf)(t) = \Delta(t)^{\frac{1}{2}} f(t^{-1}) \; ; \; (Jf)(t) = \Delta(t)^{\frac{1}{2}} \overline{f(t^{-1})} .$$

$\hat{G}$: The dual group of an abelian locally compact group G or the set of (unitary) equivalence classes of irreducible (continuous) unitary representations of G on ℓ^2 spaces.

$\chi_p, \chi_q, \ldots$: Normalized characters of G corresponding to $p, q, \ldots \in \hat{G}$.

$\{\pi, \mathfrak{H}_\pi\}, \ldots$: Unitary representations π of G on $\mathfrak{H}_\pi$.

$\{\bar{\pi}, \mathfrak{H}_{\bar{\pi}}\}$ = The unitary representation conjugate to $\{\pi, \mathfrak{H}_\pi\}$.

$\mathcal{R}$ = The ring of unitary representations of G, (Definition I.3.1).

$\mathrm{End}(\mathfrak{N})$ = The set of endomorphisms of $\mathfrak{N}$.

$\{\rho, \eta\}$ = The Roberts actions of $\mathcal{R}$, (Definition I.3.2).

$\mathcal{J}_G(\pi_1, \pi_2)$ = The set of intertwiners of π_1 and π_2 in $\mathcal{R}$.

$\mathcal{J}_G(\rho_1, \rho_2)$ = The set of intertwiners of ρ_1 and ρ_2 in $\mathrm{End}(\mathfrak{N})$.

$\mathcal{H}_\alpha(\mathfrak{M})$ = The set of all Hilbert spaces $\mathfrak{K}$ in $\mathfrak{M}$ such that $\alpha_t(\mathfrak{K}) = \mathfrak{K}$. for all t.

$\rho_{\mathfrak{K}}$ = The endomorphism of $\mathfrak{N}$: $\rho_{\mathfrak{K}}(a)x = xa$ for $x \in \mathfrak{K}$ and $a \in \mathfrak{N}$.

$\mathfrak{n} \times_\rho \mathfrak{R}$ = The crossed product of $\mathfrak{n}$ by $\mathfrak{R}$ with respect to $\{\rho,\eta\}$, (Definition IV.4.3).
$\mathcal{E}_\alpha$ = The $\mathfrak{m}^\alpha$-valued weight on $\mathfrak{m}$ given by $\langle \mathcal{E}_\alpha(x),\omega\rangle = \langle \alpha(x),\omega \otimes \mu_G'\rangle$.
$\mathcal{E}_\delta$ = The $\mathfrak{n}^\delta$-valued weight on $\mathfrak{n}$ given by $\langle \mathcal{E}_\delta(y),\omega\rangle = \langle \delta(y),\omega \otimes \psi_G\rangle$.
$\mathfrak{q}_\alpha = \{x \in \mathfrak{m} : \mathcal{E}_\alpha(x^*x)$ exists$\}$, $\mathfrak{p}_\alpha = \mathfrak{q}_\alpha^*\mathfrak{q}_\alpha$.
$\mathfrak{q}_\delta = \{y \in \mathfrak{n} : \mathcal{E}_\delta(y^*y)$ exists$\}$, $\mathfrak{p}_\delta = \mathfrak{q}_\delta^*\mathfrak{q}_\delta$.
Ker $\alpha \upharpoonright \mathcal{C}_\mathfrak{m} = \{t \in G : \alpha_t = 1$ on $\mathcal{C}_\mathfrak{m}\}$
H: A closed subgroup of G.
$\mathfrak{m} \times_\alpha H = \alpha(\mathfrak{m}) \vee (\mathbf{C} \otimes \rho(H)'')$.
$\mathcal{L}^\infty(H\backslash G) = L^\infty(G) \cap \lambda(H)'$.
$\mathcal{L}^\infty(G/H) = L^\infty(G) \cap \rho(H)'$.
$\mathfrak{n} \times_\delta (H\backslash G) = \delta(\mathfrak{n}) \vee (\mathbf{C} \otimes \mathcal{L}^\infty(H\backslash G))$.

CHAPTER I.

ACTION, CO-ACTION AND DUALITY.

Introduction. This chapter is devoted to the formulation of the duality for crossed products of von Neumann algebras involving non-commutative automorphism groups. To do this, we must reformulate the duality theorem for abelian automorphism groups. In §1, the usual duality theorem for crossed products involving only abelian groups is presented without proof together with its consequence in the structure theory of von Neumann algebras of type III. In §2, we shall review first the duality principle for abelian groups to pave the way for noncommutative groups. We then formulate the duality theorem for crossed products incolving non-commutative groups. We present a proof which takes care of the both cases, abelian and non-abelian. Here we take the Hopf-von Neumann algebra approach to the duality principle. Namely, showing that a continuous action α of a locally compact group G on a von Neumann algebra $\mathfrak{M}$ corresponds uniquely to an isomorphism π of $\mathfrak{M}$ into $\mathfrak{M} \bar{\otimes} L^\infty(G)$ such that $(\pi \otimes \iota) \circ \pi = (\iota \otimes \alpha_G) \circ \pi$, where α_G is the isomorphism of $L^\infty(G)$ into $L^\infty(G) \bar{\otimes} L^\infty(G)$ given by $(\alpha_G f)(s,t) = f(st)$, we introduce a co-action δ of G on $\mathfrak{M}$ as an isomorphism of $\mathfrak{M}$ into $\mathfrak{M} \bar{\otimes} \mathcal{R}(G)$ such that $(\delta \otimes \iota) \circ \delta = (\iota \otimes \delta_G) \circ \delta$, where $\mathcal{R}(G)$ is the von Neumann algebra generated by the right regular representation ρ of G and δ_G is the isomorphism of $\mathcal{R}(G)$ into $\mathcal{R}(G) \bar{\otimes} \mathcal{R}(G)$ such that $\delta_G(\rho(s)) = \rho(s) \otimes \rho(s)$, $s \in G$. We then prove the duality theorem, Theorems 2.5 and 2.7, as the non-commutative version of the usual duality theorem mentioned above.

Section 3 is devoted to another approach to the duality principle due to Roberts which follows more closely the spirit of the Tannaka-Tatsuuma duality theorem referring directly to a ring of representations. Here, the notion of Hilbert spaces in a von Neumann algebra plays a crucial role, which is a replacement of unitaries in the case of abelian groups. We shall see in the subsequent sections that the Hopf-von Neumann algebra approach is convenient in constructing the crossed product while the Roberts approach has an advantage in the analysis of automorphism groups over the former one. Section 4 is merely for convenience of the reader.

§1. Duality for crossed products. (Abelian case)

We begin with discussion of a duality for crossed products involving only abelian groups first.

Let G be a locally compact abelian group with a Haar measure ds, where we use the additive symbol for the group product. We denote by $\hat{G}$ the dual group of G with the Placherel measure dp. Given an action[1] α of G on a von Neumann algebra $\{\mathfrak{M},\mathfrak{H}\}$, the crossed product of $\mathfrak{M}$ by α, which will be denoted by $\mathfrak{M}\times_\alpha G$, is constructed as follows: A representation π_α of $\mathfrak{M}$ on $\mathfrak{H}\otimes L^2(G)$ is given by

(1.1) $$\pi_\alpha(x)\xi(t) = \alpha_t(x)\xi(t),\ x\in\mathfrak{M},\quad t\in G,\quad \xi\in\mathfrak{H}\otimes L^2(G)\ ;$$

a unitary representation u of G on $\mathfrak{H}\otimes L^2(G)$ is then given by:

(1.2) $$u(s)\xi(t) = \xi(s+t),\ s,t\in G\ .$$

Then $\mathfrak{M}\times_\alpha G$ is the von Neumann algebra on $\mathfrak{H}\otimes L^2(G)$ generated by $\pi_\alpha(\mathfrak{M})$ and $u(G)$. Next, we construct a unitary representation v of $\hat{G}$ on $\mathfrak{H}\otimes L^2(G)$ by the following

(1.3) $$(v(p)\xi)(s) = \overline{\langle s,p\rangle}\,\xi(s),\quad s\in G,\quad p\in\hat{G}\ .$$

We then have

(1.4) $$\begin{aligned} v(p)\pi_\alpha(x)v(p)^* &= \pi_\alpha(x),\quad x\in\mathfrak{M},\quad p\in\hat{G}\ ;\\ v(p)u(s)v(p)^* &= \langle s,p\rangle u(s),\ s\in G,\ p\in\hat{G}\ .\end{aligned}$$

Hence the automorphism of $\mathfrak{L}(\mathfrak{H}\otimes L^2(G))$ induced by $v(p)$ leaves the generators of $\mathfrak{M}\times_\alpha G$ invariant up to multiple by scalars, so that it gives rise to an automorphism of $\mathfrak{M}\times_\alpha G$, which will be denoted by $\hat{\alpha}_p$. Thus we obtain an action $\hat{\alpha}$ of $\hat{G}$ on $\mathfrak{M}\times_\alpha G$. We shall call it the dual action.

Theorem 1.1 In the above situation,

$$(\mathfrak{M}\times_\alpha G)\times_{\hat{\alpha}}\hat{G}\cong\mathfrak{M}\mathbin{\bar{\otimes}}\mathfrak{L}(L^2(G))\ .$$

The isomorphism carries the action $\hat{\hat{\alpha}}$ of G on $(\mathfrak{M}\times_\alpha G)\times_{\hat{\alpha}}\hat{G}$ dual to $\hat{\alpha}$ into the action $\tilde{\alpha}$ of G on $\mathfrak{M}\mathbin{\bar{\otimes}}\mathfrak{L}(L^2(G))$ given by

---

1) An action of a locally compact group G on a von Neumann algebra $\mathfrak{M}$ means a homomorphism α of G into $\mathrm{Aut}(\mathfrak{M})$ such that $s\in G\to\alpha_s(x)\in\mathfrak{M}$ is σ-weakly continuous for each $x\in\mathfrak{M}$. The composition $\{\mathfrak{M},G,\alpha\}$ or $\{\mathfrak{M},\alpha\}$ will often be called a covariant system. The topology in $\mathrm{Aut}(\mathfrak{M})$ should however be considered as the point-norm convergence topology in the predual under the transposed action.

$$\tilde{\alpha}_s = \alpha_s \otimes \lambda_s , \quad s \in G ,$$

where λ_s means the inner automorphism of $\mathcal{L}(L^2(G))$ induced by the representation $\lambda(s)$ of G on $L^2(G)$:

$$(\lambda(s)\xi)(t) = \xi(t - s) , s,t \in G, \quad \xi \in L^2(G) .$$

The proof will be given in the next section. Applying the above theorem to the modular automorphism group, we obtain the following structure theorem for factors of type III.

<u>Theorem</u> 1.2 If $\mathfrak{M}$ is a properly infinite von Neumann algebra, then there exists a unique properly infinite but semi-finite von Neumann algebra $\mathfrak{N}$ equipped with a one parameter automorphism group $\{\theta_t\}$ and a faithful, semi-finite, normal trace τ such that $\tau \circ \theta_t = e^{-t}\tau$ and $\mathfrak{M} \cong \mathfrak{N} \times_\theta \mathbb{R}$. If $\mathfrak{M}$ is a factor, then θ is ergodic on the center $\mathcal{C}_\mathfrak{N}$ of $\mathfrak{N}$. If $\mathfrak{M}$ is of type III, then $\{\mathcal{C}_\mathfrak{N}, \theta\}$ does not admit a multiple of $L^\infty(\mathbb{R})$ with translation as a direct summand and $\mathfrak{N}$ must be of type II_∞.

We leave the detail to the original paper [69] and subsequent structure analysis [14].

<u>NOTES</u>

Historically, the duality theorem for crossed product, Theorem 1.1, was discovered through the structure analysis of a factor of type III, Theorem 1.2, [14, 69]. Related references: [2,12,14,67,68,69].

§2. Duality for crossed products. (General case)

In this section, we shall discuss a general duality theorem for crossed products involving non-abelian groups. Since we do not have a dual group for a non-abelian group, we must reformulate the duality theorem for abelian groups without making use of the dual group before we move to the non-abelian case.

Suppose G is an abelian locally compact group. Let $\mathcal{R}(G)$ denote the von Neumann algebra on $L^2(G)$ generated by the regular representation ρ of G, where

$$\rho(s)\xi(t) = \xi(s + t) \ , \quad s,t \in G \ , \ \xi \in L^2(G) \ .$$

Denoting by $\mathcal{F}$ the Fourier transform of $L^2(G)$ onto $L^2(\hat{G})$, we have

$$\mathcal{F}L^\infty(G)\mathcal{F}^{-1} = \mathcal{R}(\hat{G}), \ \mathcal{F}\mathcal{R}(G)\mathcal{F}^{-1} = L^\infty(\hat{G}) \ ,$$

where we consider $L^\infty(G)$ (resp. $L^\infty(\hat{G})$) as a von Neumann algebra on $L^2(G)$ (resp. $L^2(\hat{G})$) acting by multiplication. Since we want to eliminate $\hat{G}$, we identify $L^\infty(\hat{G})$ with $\mathcal{R}(G)$ via the Fourier transform. We then formulate the duality of G and $\hat{G}$ in terms of $L^\infty(G)$ and $\mathcal{R}(G)$. Indeed, the Hopf-von Neumann algebra approach tells us that $L^\infty(G)$ and $\mathcal{R}(G)$ carry the structures dual to each other. Namely, $L^\infty(G)$ carries the co-multiplication α_G which is defined by:

$$\alpha_G(f)(s,t) = f(s + t), \ f \in L^\infty(G) \ , \quad s,t \in G \ ;$$

and $\mathcal{R}(G)$ does also the co-multiplication δ_G:

$$\delta_G(\rho(s)) = \rho(s) \otimes \rho(s) \ , \quad s \in G \ .$$

We then see that $\{L^\infty(G), \alpha_G\}$ and $\{\mathcal{R}(G), \delta_G\}$ serve as the dual system of the other. At this stage, we see that the dual group $\hat{G}$ does not appear in an explicit form.

Let us review what we have done in the above. Indeed, the above procedure means that we translated the group structure of G into the von Neumann algebra $L^\infty(G)$ together with the isomorphism α_G, the co-multiplication, of $L^\infty(G)$ into $L^\infty(G) \overline{\otimes} L^\infty(G)$; and that of the dual group $\hat{G}$ into $\mathcal{R}(G)$ with the isomorphism δ_G of $\mathcal{R}(G)$ into $\mathcal{R}(G) \overline{\otimes} \mathcal{R}(G)$. The both systems $\{L^\infty(G),\alpha_G\}$ and $\{\mathcal{R}(G),\delta_G\}$ satisfy the same commutative diagram:

$$\begin{array}{ccc} \mathfrak{M} & \xrightarrow{\ \pi\ } & \mathfrak{M} \overline{\otimes} \mathfrak{M} \\ \downarrow{\scriptstyle \pi} & & \downarrow{\scriptstyle \pi \otimes \iota} \\ \mathfrak{M} \overline{\otimes} \mathfrak{M} & \xrightarrow[\ \iota \otimes \pi\ ]{} & \mathfrak{M} \overline{\otimes} \mathfrak{M} \overline{\otimes} \mathfrak{M} \end{array} \ .$$

We now remove the commutativity assumption from G, i.e. G is now a locally compact (not necessarily abelian) group with a right Haar measure ds. Let ρ (resp. λ) be the right (resp. left) regular representation of G on $L^2(G)$, and let $\mathcal{R}(G) = \rho(G)''$ (resp. $\mathcal{R}'(G) = \lambda(G)''$). We define unitary operators V_G and W_G on $L^2(G) \otimes L^2(G) = L^2(G \times G)$ as follows:

$$(2.1) \qquad \begin{cases} V_G\xi(s,t) = \xi(st,t), \ \xi \in L^2(G \times G), \ s,t \in G \ ; \\ \\ W_G\xi(s,t) = \xi(s,ts) \ . \end{cases}$$

For each $f \in L^\infty(G)$ and $x \in \mathcal{R}(G)$, set

$$(2.2) \qquad \begin{cases} \alpha_G(f) = V_G(f \otimes 1)V_G^* = \mathrm{Ad}_{V_G}(f \otimes 1) \ ; \\ \\ \delta_G(x) = W_G^*(x \otimes 1)W_G = \mathrm{Ad}_{W_G^*}(x \otimes 1) \ . \end{cases}$$

We then have

$$(2.3) \qquad \begin{cases} (\alpha_G \otimes \iota) \cdot \alpha_G = (\iota \otimes \alpha_G) \cdot \alpha_G \ ; \\ \\ (\delta_G \otimes \iota) \cdot \delta_G = (\iota \otimes \delta_G) \cdot \delta_G \ . \end{cases}$$

We must now formulate the notion of an action of G on a von Neumann algebra $\mathfrak{M}$ in terms of $\{L^\infty(G), \alpha_G\}$. Suppose that an action α of G on $\mathfrak{M}$ in the conventional sense is given. This means that we have a map : $(x,s) \in \mathfrak{M} \times G \to \alpha_s(x) \in \mathfrak{M}$ with certain properties. If we fix an $x \in \mathfrak{M}$, then we get an $\mathfrak{M}$-valued function: $s \in G \to \alpha_s(x) \in \mathfrak{M}$ on G, which is, in turn, an element $\pi_\alpha(x)$ of $\mathfrak{M} \bar{\otimes} L^\infty(G)$. Namely, we have a map $\pi_\alpha : x \in \mathfrak{M} \to \pi_\alpha(x) \in \mathfrak{M} \bar{\otimes} L^\infty(G)$ from $\mathfrak{M}$ into $\mathfrak{M} \bar{\otimes} L^\infty(G)$. The isomorphism property of each α_s reflects to the isomorphism property of π_α, i.e. π_α is an isomorphism of $\mathfrak{M}$ into $\mathfrak{M} \bar{\otimes} L^\infty(G)$. The homomorphism property of the map: $s \in G \to \alpha_s \in \mathrm{Aut}(\mathfrak{M})$ is translated to the commutativity of the diagram:

$$\begin{array}{ccc} \mathfrak{M} & \xrightarrow{\ \pi_\alpha\ } & \mathfrak{M} \bar{\otimes} L^\infty(G) \\ \Big\downarrow \pi_\alpha & & \Big\downarrow (\iota \otimes \alpha_G) \\ \mathfrak{M} \bar{\otimes} L^\infty(G) & \xrightarrow[(\pi_\alpha \otimes \iota)]{} & \mathfrak{M} \bar{\otimes} L^\infty(G) \bar{\otimes} L^\infty(G) \ . \end{array}$$

We then discover that the isomorphism π_α has already appeared in the construction of the crossed product as in (1.1). In order to complete or to start our program, we must, however, prove the following first:

<u>Proposition</u> 2.1. A normal[2] isomorphism π of $\mathfrak{M}$ into $\mathfrak{M} \bar{\otimes} L^\infty(G)$ satisfying the equality:

(2.4)
$$(\pi \otimes \iota) \circ \pi = (\iota \otimes \alpha_G) \circ \pi$$

gives rise to an action α of G on $\mathfrak{M}$ with $\pi_\alpha = \pi$.

<u>Proof</u>. First we shall prove that $\pi(\mathfrak{M})$ is invariant under $\{\iota \otimes \rho_t ; t \in G\}$. Making use of the duality of $\mathfrak{M}$ and $\mathfrak{M}_*$, we define a linear transformation π_f on $\mathfrak{M}$ for each $f \in L^1(G)$ as follows:

$$\langle \pi_f(x), \omega \rangle = \langle \pi(x), \omega \otimes f \rangle , \quad x \in \mathfrak{M}, \ \omega \in \mathfrak{M}_* .$$

Noticing that for each $f, g \in L^1(G), h \in L^\infty(G)$,

$$\langle h, f*g \rangle = \langle \alpha_G(h), f \otimes g \rangle = \langle \rho_g(h), f \rangle^{3)} ,$$

we have

$$\begin{aligned}\langle \pi \circ \pi_f(x), \omega \otimes g \rangle &= \langle \pi_f(x), \pi_*(\omega \otimes g) \rangle \\ &= \langle \pi(x), \pi_*(\omega \otimes g) \otimes f \rangle = \langle (\pi \otimes \iota) \circ \pi(x), \omega \otimes g \otimes f \rangle \\ &= \langle (\iota \otimes \alpha_G) \circ \pi(x), \omega \otimes g \otimes f \rangle \\ &= \langle (\iota \otimes \rho_f) \circ \pi(x), \omega \otimes g \rangle ;\end{aligned}$$

hence we get $\pi \circ \pi_f = (\iota \otimes \rho_f) \circ \pi$, $f \in L^1(G)$. Thus, $\pi(\mathfrak{M})$ is globally invariant under $\{\iota \otimes \rho_f : f \in L^1(G)\}$. If we choose a net $\{f_i\}$ in $L^1(G)$ such that $f_i(s)ds$ converges to the Dirac measure at a given point $s \in G$, then $\{\rho_{f_i}\}$ converges to ρ_s in an appropriate sense, so that $\pi(\mathfrak{M})$ is invariant under $\iota \otimes \rho_s$ for $s \in G$. We then define $\alpha_s = \pi^{-1} \circ (\iota \otimes \rho_s) \circ \pi$, $s \in G$. Since we have $\alpha_f = \pi_f$, $f \in L^1(G)$, we get $\pi = \pi_\alpha$. Q.E.D.

Therefore, we have established that an action α of G on $\mathfrak{M}$ may be identified with an isomorphism π_α of $\mathfrak{M}$ into $\mathfrak{M} \bar{\otimes} L^\infty(G)$ satisfying (2.4). We shall use the same symbol α for an action and the isomorphism π_α, and call it an <u>action</u> of G on $\mathfrak{M}$. In this respect, the co-multiplication α_G is indeed an action of G on $L^\infty(G)$ induced by ρ.

<u>Definition</u> 2.2. The von Neumann algebra generated by $\alpha(\mathfrak{M})$ and $\mathbf{C} \otimes \mathcal{R}(G)$ is called the <u>crossed product</u> of $\mathfrak{M}$ by G with respect to α (or simply the crossed product of $\mathfrak{M}$ by α), which will be denoted by $\mathfrak{M} \times_\alpha G$.

---

2) We consider throughout only normal maps for von Neumann algebras.

3) $\rho_g = \int_G g(t)\rho_t \, dt$.

We are now ready to define an action of the "dual" of G on a von Neumann algebra $\mathfrak{N}$. We should simply replace $L^\infty(G)$ by $\mathcal{R}(G)$ and α_G by δ_G in the commutative diagram for π_α. Namely, we have the following:

Definition 2.3. A co-action of G on $\mathfrak{N}$ is an isomorphism δ of $\mathfrak{N}$ into $\mathfrak{N} \bar{\otimes} \mathcal{R}(G)$ such that

$$(2.5) \qquad (\delta \otimes \iota) \circ \delta = (\iota \otimes \delta_G) \circ \delta .$$

The von Neumann algebra generated by $\delta(\mathfrak{N})$ and $\mathbf{C} \otimes L^\infty(G)$ is called the crossed product of $\mathfrak{N}$ by G with respect to δ (or simply the crossed product of $\mathfrak{N}$ by δ), which will be denoted by $\mathfrak{N} \times_\delta G$.

Proposition 2.4. (Dual co-action and action). i) Given an action α of G on $\mathfrak{M}$, if we set

$$(2.6) \qquad \hat{\alpha}(y) = \mathrm{Ad}_{(1 \otimes W_G^*)}(y \otimes 1), \quad y \in \mathfrak{M} \times_\alpha G ,$$

then $\hat{\alpha}$ is a co-action of G on $\mathfrak{M} \times_\alpha G$, which will be called dual to α.

ii) Given a co-action δ of G on $\mathfrak{N}$ first, if we set

$$(2.7) \qquad \hat{\delta}(x) = \mathrm{Ad}_{(1 \otimes V_G')}(x \otimes 1), \quad x \in \mathfrak{N} \times_\delta G ,$$ [4]

then $\hat{\delta}$ is an action of G on $\mathfrak{N} \times_\delta G$, which will be called dual to δ.

Proof i) Since $W_G \in L^\infty(G) \bar{\otimes} \mathcal{R}(G)$, $1 \otimes W_G^*$ and $\alpha(x) \otimes 1$, $x \in \mathfrak{M}$, commute, so that

$$(2.8) \qquad \hat{\alpha}(\alpha(x)) = \alpha(x) \otimes 1 \in (\mathfrak{M} \times_\alpha G) \otimes \mathbf{C} .$$

Since we have

$$W_G^*(\rho(r) \otimes 1)W_G = \rho(r) \otimes \rho(r), \quad r \in G ,$$

we get

$$(2.9) \qquad \hat{\alpha}(1 \otimes \rho(r)) = 1 \otimes \rho(r) \otimes \rho(r) \in (\mathfrak{M} \times_\alpha G) \bar{\otimes} \mathcal{R}(G) .$$

Equality (2.5) is also seen by checking the generators $\alpha(\mathfrak{M})$ and $\rho(G)$ of $\mathfrak{M} \times_\alpha G$.

ii) We have $V_G' \in \mathcal{R}'(G) \bar{\otimes} L^\infty(G)$, so that $\delta(x) \otimes 1$, $x \in \mathfrak{N}$, and $1 \otimes V_G'$ commute; hence

$$(2.10) \qquad \hat{\delta} \circ \delta(x) = \delta(x) \otimes 1, \quad x \in \mathfrak{N} .$$

For each $f \in L^\infty(G)$, we have

$$\lambda(f) = V_G'(f \otimes 1)V_G'^* ,$$

---

4) For the definition of V_G', see the list of symbols.

where $\lambda(f)(s,t) = f(t^{-1}s)$, so that

$$\hat{\delta}(1 \otimes f) = 1 \otimes \lambda(f) . \tag{2.11}$$

But λ is nothing but π_λ with the notation in Proposition 2.1. Thus, δ maps $\mathfrak{n} \times_\delta G$ into $(\mathfrak{n} \times_\delta G) \bar{\otimes} L^\infty(G)$. Equality (2.4) can be checked by looking at the generators separately.
Q.E.D.

<u>Theorem</u> 2.5. (Duality for actions). If α is an action of G on $\mathfrak{M}$, then

i) $$(\mathfrak{M} \times_\alpha G) \times_{\hat{\alpha}} G \cong \mathfrak{M} \bar{\otimes} \mathfrak{L}(L^2(G)) ;$$

ii) the second dual action $\hat{\hat{\alpha}}$ of G on $(\mathfrak{M} \times_\alpha G) \times_{\hat{\alpha}} G$ is conjugate to the action $\tilde{\alpha}$ of G on $\mathfrak{M} \bar{\otimes} \mathfrak{L}(L^2(G))$ defined by $\tilde{\alpha}_t = \alpha_t \otimes \lambda_t$ under the above isomorphism.

<u>Lemma</u> 2.6. $\mathfrak{M} \bar{\otimes} L^\infty(G) = \alpha(\mathfrak{M}) \vee (\mathbf{C} \otimes L^\infty(G))$.

<u>Proof</u>. Let $C_\infty(G,\mathfrak{M})$ denote the C^*-algebra of all continuous $\mathfrak{M}$-valued functions on G vanishing at infinity with respect to the norm topology in $\mathfrak{M}$. Of course, $C_\infty(G,\mathfrak{M})$ is naturally identified with the C^*-tensor product $C_\infty(G) \hat{\otimes}_* \mathfrak{M}$. For any two distinct points s,t in G and $x,y \in \mathfrak{M}$, we choose two continuous functions f and g with compact support such that $f(s) = g(t) = 1$ and $f(t) = g(s) = 0$. Let $a = \alpha_s^{-1}(x)$ and $b = \alpha_t^{-1}(y)$. Set $\alpha(a)(1 \otimes f) + \alpha(b)(1 \otimes g) = z \in \alpha(\mathfrak{M}) \vee (\mathbf{C} \otimes L^\infty(G))$. We then have

$$z(s) = \alpha_s(a)f(s) + \alpha_s(b)g(s) = \alpha_s(a) = x ;$$

$$z(t) = y .$$

Therefore, the partition of unity shows that every element of $C_\infty(G,\mathfrak{M})$ is well-approximated by $\alpha(\mathfrak{M}) \vee (\mathbf{C} \otimes L^\infty(G))$. Thus, our assertion follows. Q.E.D.

<u>Proof of Theorem</u> 2.5. By the previous lemma, $\mathfrak{M} \bar{\otimes} \mathfrak{L}(L^2(G))$ is generated by $\alpha(\mathfrak{M})$, $\mathbf{C} \otimes \mathfrak{R}(G)$ and $\mathbf{C} \otimes L^\infty(G)$. By definition, $(\mathfrak{M} \times_\alpha G) \times_{\hat{\alpha}} G$ is generated by $\hat{\alpha} \circ \alpha(\mathfrak{M}) = \alpha(\mathfrak{M}) \otimes \mathbf{C}$, $\hat{\alpha}(\mathbf{C} \otimes \mathfrak{R}(G)) = \mathbf{C} \otimes \delta_G(\mathfrak{R}(G))$ and $\mathbf{C} \otimes \mathbf{C} \otimes L^\infty(G)$. In the algebra $\mathfrak{M} \bar{\otimes} \mathfrak{L}(L^2(G)) \bar{\otimes} \mathfrak{L}(L^2(G))$, we set

$$\pi(x) = (1 \otimes V_G)^*(\alpha \otimes \iota)(x)(1 \otimes V_G), \ x \in \mathfrak{M} \bar{\otimes} \mathfrak{L}(L^2(G)) . \tag{2.12}$$

We then have

$$\begin{cases} \pi(\alpha(x)) = \alpha(x) \otimes 1 , & x \in \mathfrak{M} ; \\ \pi(1 \otimes \rho(r)) = 1 \otimes \rho(r) \otimes \rho(r) , & r \in G ; \\ \pi(1 \otimes f) = 1 \otimes 1 \otimes f , & f \in L^\infty(G) . \end{cases} \tag{2.13}$$

Thus, we get $\pi(\mathfrak{m} \bar{\otimes} \mathcal{L}(L^2(G))) = (\mathfrak{m} \times_\alpha G) \times_{\hat{\alpha}} G$. Since $(\alpha_s \otimes \lambda_s) \circ \alpha = \alpha$ for each $s \in G$, we have

$$(2.14) \quad \begin{cases} \tilde{\alpha} \circ \alpha(x) = \alpha(x) \otimes 1 , & x \in \mathfrak{m} ; \\ \tilde{\alpha}(1 \otimes \rho(r)) = 1 \otimes \rho(r) \otimes 1 , & r \in G ; \\ \tilde{\alpha}(1 \otimes f) = 1 \otimes \lambda(f) , & f \in L^\infty(G) . \end{cases}$$

Applying $\pi \otimes \iota$ to the above, we see that π intertwines $\tilde{\alpha}$ and $\hat{\hat{\alpha}}$. Q.E.D.

Proof of Theorem 1.1. Suppose G is abelian. The discussion at the beginning of this section together with Proposition 2.1 and Theorem 2.5 yields the desired conclusion. Q.E.D.

In order to put Theorem 2.5 in the form symmetric to the next result, we express the action $\tilde{\alpha}$ of G on $\mathfrak{m} \bar{\otimes} \mathcal{L}(L^2(G))$ in the following formula, which is directly checked by looking at the generators:

$$(2.15) \quad \tilde{\alpha}(x) = \mathrm{Ad}_{(1 \otimes V_G')} \circ (\iota \otimes \sigma) \circ (\alpha \otimes \iota)(x), \ x \in \mathfrak{m} \bar{\otimes} \mathcal{L}(L^2(G)).$$

Theorem 2.7. (Duality for co-actions) If δ is a co-action of G on $\mathfrak{n}$, then

$$(2.16) \quad \tilde{\delta}(y) = \mathrm{Ad}_{(1 \otimes W_G)} \circ (\iota \otimes \sigma) \circ (\delta \otimes \iota)(y), \quad y \in \mathfrak{n} \bar{\otimes} \mathcal{L}(L^2(G)) .$$

is a co-action of G on $\mathfrak{n} \bar{\otimes} \mathcal{L}(L^2(G))$ and

$$\{(\mathfrak{n} \times_\delta G) \times_{\hat{\delta}} G, \hat{\hat{\delta}}\} \cong \{\mathfrak{n} \bar{\otimes} \mathcal{L}(L^2(G)), \tilde{\delta}\} .$$

The map δ defined by

$$\delta(y) = \mathrm{Ad}_{1 \otimes W_G^*}(y \otimes 1) , \quad y \in \mathcal{L}(\mathfrak{K} \otimes L^2(G))$$

is a co-action of G on $\mathcal{L}(\mathfrak{K} \otimes L^2(G))$, whose restriction to $\mathcal{R}(G)$ agrees with the co-action δ_G. We shall use this δ in the following lemma.

Lemma 2.8. (i) If $\mathfrak{J}$ is the set of all compact subsets of G ordered by set inclusion, then there is a net $\{\varphi_K : K \in \mathfrak{J}\}$ in $A(G) \cap \mathcal{K}(G)$ such that $\varphi_K(t)$ converges to 1 for each $t \in G$.

(ii) If $y_j \in \mathcal{L}(\mathfrak{K} \otimes L^2(G))$ and $\xi_j \in \mathcal{K}(G,\mathfrak{K})$ for $j = 1,2$, then for any $\varepsilon > 0$ and $\eta_j \in \mathfrak{K} \otimes L^2(G)$ there exists a $\psi \in A(G) \cap \mathcal{K}(G)$ such that

$$|((\delta_\psi(y_j) - y_j)\xi_j \mid \eta_j)| < \varepsilon ,$$

where δ_ψ is defined by $\langle \delta_\psi(y), \omega \rangle = \langle \delta(y), \omega \otimes \psi \rangle$ for $\omega \in \mathcal{L}(\mathfrak{K} \otimes L^2(G))_*$.

Proof. (i) Let f_o be an element in $\mathcal{K}(G)$ with

$$f_o(t) \geq 0 \quad \text{and} \quad \int f_o(s)ds = 1 .$$

For each $K \in \mathfrak{F}$ we set

$$f_K(t) = \int_K f_o(st)ds \quad \text{and} \quad \varphi_K = \omega_{f_o, f_K} .$$

Then $0 \leq f_K(t) \leq 1$ and $\varphi_K \in A(G) \cap \mathcal{K}(G)$. Since $f_K(t)$ converges to 1 as K tends to G, $\varphi_K(t)$ converges to 1.

(ii) The functions $(1 \otimes W_G)(\xi_j \otimes f_o)$ for $j = 1,2$ are elements in $\mathcal{K}(G \times G, \mathfrak{K})$ supported by $K_1 \times K_2$ for some $K_i \in \mathfrak{F}$. If $h \in \mathcal{K}(G)_+$ satisfies $h = 1$ on K_2, then

$$(\delta_{\varphi_K}(y_j)\xi_j \mid \eta_j) = ((y_j \otimes 1)(1 \otimes W_G)(\xi_j \otimes f_o) \mid (1 \otimes W_G)(\eta_j \otimes f_K))$$

$$= ((y_j \otimes 1)(1 \otimes W_G)(\xi_j \otimes f_o) \mid (1 \otimes 1 \otimes h)(1 \otimes W_G)(\eta_j \otimes f_K))$$

and hence

$$|(\delta_{\varphi_K}(y_j)\xi_j|\eta_j)| \leq \|(y_j \otimes 1)(1 \otimes W_G)(\xi_j \otimes f_o)\| \; \|h\| \; \|\eta_j\| ,$$

where the right hand side are L^2-norms. Let $\mathfrak{L}_o$ be the set of all finite linear combinations of elements in $\mathfrak{L}(\mathfrak{K}) \otimes C$, $C \otimes L^\infty(G)$ and $C \otimes \mathcal{R}(G)$. Since $f\rho(s)g\rho(t) = f\rho_s(g)\rho(st)$ and $(f\rho(s))^* = \rho_{s^{-1}}(f)\rho(s)^*$, the set $\mathfrak{L}_o$ is a strongly dense $*$ sub-algebra of $\mathfrak{L}(\mathfrak{K} \otimes L^2(G))$. For any $\varepsilon > 0$ there exists $x_j \in \mathfrak{L}_o$ such that

$$|((y_j - x_j)\xi_j \mid \eta_j)| < \varepsilon \quad \text{and} \quad \|((y_j - x_j) \otimes 1)(1 \otimes W_G)(\xi_j \otimes f_o)\| < \varepsilon$$

for $j = 1,2$. Therefore

$$|((\delta_{\varphi_K}(y_j) - y_j)\xi_j|\eta_j)|$$

$$\leq |(\delta_{\varphi_K}(y_j - x_j)\xi_j|\eta_j)| + |((\delta_{\varphi_K}(x_j) - x_j)\xi_j \mid \eta_j)| + |((x_j - y_j)\xi_j \mid \eta_j)|$$

$$\leq \varepsilon\|h\| \; \|\eta_j\| + |((\delta_{\varphi_K}(x_j) - x_j)\xi_j \mid \eta_j)| + \varepsilon .$$

Thus it remains to show that the second term converges to 0. Since $\delta(z_o) = \mathrm{Ad}_{1 \otimes W_G}(z_o \otimes 1) = z_o \otimes \rho(r)$ for $z_o = y \otimes f\rho(r)$, it follows that, for any $z \in \mathfrak{L}_o$ of the form $\sum_{k=1}^n z_k$ with $z_k = y_k \otimes f_k\rho(r_k)$,

$$(\delta_{\varphi_K}(z)\xi_j \mid \eta_j) = \sum_{k=1}^{n} (z_k\xi_j \mid \eta_j)\varphi_K(r_k) \to (z\xi_j \mid \eta_j) . \qquad \text{Q.E.D.}$$

<u>Lemma</u> 2.9. (i) If $\delta(x) = \mathrm{Ad}_{1\otimes W_G^*}(x \otimes 1)$ for $x \in \mathcal{L}(\mathfrak{K} \otimes L^2(G))$, then

$$x \in \{\delta_\varphi(x) : \varphi \in A(G) \cap \mathcal{K}(G)\}'' .$$

(ii) If δ is a co-action of G on $\mathcal{M}$, then the set of all $\delta_\varphi(y)$ with $y \in \mathcal{M}$ and $\varphi \in A(G) \cap \mathcal{K}(G)$ defined by $\langle \delta_\varphi(y), \omega\rangle = \langle \delta(y), \omega \otimes \varphi\rangle$ for $\omega \in \mathcal{M}_*$ is σ-weakly dense in $\mathcal{M}$.

<u>Proof</u>. (i) Let $\xi, \eta \in \mathfrak{K}$ and $f, g \in \mathcal{K}(G)$. For any $z \in \mathcal{L}(\mathfrak{K} \otimes L^2(G))$ and any $\varepsilon > 0$ there exists a $\psi \in A(G) \cap \mathcal{K}(G)$ such that

$$|((\delta_\psi(x) - x)(\xi \otimes f) | z^*(\eta \otimes g))| < \varepsilon$$

$$|(z(\xi \otimes f) | (\delta_\psi(x^*) - x^*)(\eta \otimes g))| < \varepsilon$$

by Lemma 2.8. If z commutes with $\delta_\varphi(x)$ for all $\varphi \in A(G) \cap \mathcal{K}(G)$, then for any $\varepsilon > 0$ we have

$$|((xz - zx)(\xi \otimes f) \mid \eta \otimes g))| < 2\varepsilon ,$$

which implies $xz = zx$.

(ii) If $y \in \mathcal{M}$, then $\delta(y) \in \mathcal{M} \bar{\otimes} \mathcal{R}(G)$. Since

$$\langle (\iota \otimes \delta_G)_\varphi(\delta(y)), \omega \otimes \psi\rangle = \langle (\iota \otimes \delta_G)(\delta(y)) , \omega \otimes \psi \otimes \varphi\rangle$$

$$= \langle (\delta \otimes \iota) \circ \delta(y), \omega \otimes \psi \otimes \varphi\rangle = \langle \delta(y), \delta_*(\omega \otimes \psi) \otimes \varphi\rangle$$

$$= \langle \delta_\varphi(y), \delta_*(\omega \otimes \psi)\rangle = \langle \delta(\delta_\varphi(y)), \omega \otimes \psi\rangle ,$$

it follows that

$$\delta(y) \in \{\delta(\delta_\varphi(y)) : \varphi \in A(G) \cap \mathcal{K}(G)\}'' .$$ Q.E.D.

<u>Lemma</u> 2.10. $\mathcal{M} \bar{\otimes} \mathcal{L}(L^2(G)) = \delta(\mathcal{M})' \vee (\mathbf{C} \otimes \mathcal{L}(L^2(G)))$.

<u>Proof</u>. It is clear that $\delta(\mathcal{M}) \vee (\mathbf{C} \otimes \mathcal{L}(L^2(G))) \subset \mathcal{M} \bar{\otimes} \mathcal{L}(L^2(G))$. To conclude the reversed inclusion, we shall show that $\{\mathcal{M}\bar{\otimes}\mathcal{L}(L^2(G))\}' = \mathcal{M}' \otimes \mathbf{C} \supset \delta(\mathcal{M}) \cap (\mathbf{C} \otimes \mathcal{L}(L^2(G)))' = \delta(\mathcal{M})' \cap (\mathcal{L}(\mathfrak{K}) \otimes \mathbf{C})$, where $\mathfrak{K}$ is the Hilbert space on which $\mathcal{M}$ acts.

For each $\varphi \in A(G)$, set

(2.18) $$\langle \delta_\varphi(y), \omega\rangle = \langle \delta(y), \omega \otimes \varphi\rangle, \ \omega \in \mathcal{M}_*, \ y \in \mathcal{M} .$$

If $x \otimes 1 \in \delta(\mathcal{M})' \cap (\mathcal{L}(\mathfrak{K}) \otimes \mathbf{C})$, then we have, for any $y \in \mathcal{M}$, $\omega \in \mathcal{L}(\mathfrak{K})_*$ and $\varphi \in A(G)$,

$$\langle x\delta_\varphi(y),\omega\rangle = \langle \delta_\varphi(y),\omega x\rangle = \langle \delta(y),\omega x\otimes\varphi\rangle$$

$$= \langle (x\otimes 1)\delta(y),\omega\otimes\varphi\rangle = \langle \delta(y)(x\otimes 1),\omega\otimes\varphi\rangle = \langle \delta(y),x\omega\otimes\varphi\rangle$$

$$= \langle \delta_\varphi(y),x\omega\rangle = \langle \delta_\varphi(y)x,\omega\rangle ,$$

so that $x\delta_\varphi(y) = \delta_\varphi(y)x$. But by Lemma 2.9.ii, $\{\delta_\varphi(y) : y \in \mathfrak{h}, \varphi \in A(G)\}$ is total in $\mathfrak{h}$; hence $x \in \mathfrak{h}'$. Q.E.D.

Proof of Theorem 2.7. Let K denote the unitary on $L^2(G)$ defined by

$$K\xi(s) = \Delta(s)^{1/2}\xi(s^{-1}) . \tag{2.19}$$

We have then

$$K\rho(s)K = \lambda(s) , \quad K\lambda(s)K = \rho(s) .$$

We define a map π of $\mathfrak{h} \mathbin{\bar{\otimes}} \mathcal{L}(L^2(G))$ into $\mathfrak{h} \mathbin{\bar{\otimes}} \mathcal{L}(L^2(G)) \mathbin{\bar{\otimes}} \mathcal{L}(L^2(G))$ as follows:

$$\pi(x) = (1\otimes 1\otimes K)(1\otimes W_G)(\delta\otimes\iota)(x)(1\otimes W_G^*)(1\otimes 1\otimes K) . \tag{2.20}$$

If $y \in \mathfrak{h}$ and $x = \delta(y)$, then $(\delta\otimes\iota)(x) = (\iota\otimes\delta_G)\circ\delta(y)$ by (2.5), hence (2.2) entails that

$$\pi(\delta(y)) = (1\otimes 1\otimes K)(\delta(y)\otimes 1)(1\otimes 1\otimes K) = \delta(y)\otimes 1 . \tag{2.21}$$

Then, by direct computation, we have

$$\pi(1\otimes f) = 1\otimes\lambda(f) , \quad \pi(1\otimes\lambda(r)) = 1\otimes 1\otimes\rho(r) . \tag{2.22}$$

By the previous lemma, $\mathfrak{h} \mathbin{\bar{\otimes}} \mathcal{L}(L^2(G))$ is generated by $\delta(\mathfrak{h})$, $\mathbf{C}\otimes L^\infty(G)$ and $\mathbf{C}\otimes\mathcal{R}'(G)$; thus π maps the generators of $\mathfrak{h} \mathbin{\bar{\otimes}} \mathcal{L}(L^2(G))$ onto those of $(\mathfrak{h}\times_\delta G)\times_{\hat{\delta}} G$ by Proposition 2.4.ii.

It remains to be shown that $\hat{\hat{\delta}}\circ\pi = (\pi\otimes\iota)\circ\tilde{\delta}$. From (2.8), it follows that

$$\hat{\hat{\delta}}\circ\hat{\delta}(x) = \hat{\delta}(x)\otimes 1 , \quad x\in(\mathfrak{h}\times_\delta G) ;$$

and trivially

$$\hat{\hat{\delta}}(1\otimes 1\otimes\rho(r)) = 1\otimes 1\otimes\rho(r)\otimes\rho(r) .$$

On the other hand, we have, by direct computation,

$$\begin{cases} \tilde{\delta}(\delta(y)) = \delta(y)\otimes 1 , & y\in\mathfrak{h} ; \\ \tilde{\delta}(1\otimes f) = 1\otimes f\otimes 1 , & f\in L^\infty(G) ; \\ \tilde{\delta}(1\otimes\lambda(r)) = 1\otimes\lambda(r)\otimes\rho(r) , & r\in G . \end{cases} \tag{2.23}$$

Thus, π intertwines $\tilde{\delta}$ and $\hat{\hat{\delta}}$. Q.E.D.

NOTES

The definition of a co-action, Definition 2.3, and the construction of the crossed product $\mathfrak{h} \times_\delta G$ were given independently by Landstad [42,43], Nakagami [45, 46] and Strǎtilǎ - Voiculescu - Zsidó [59,60]. Dual co-actions and dual actions, Proposition 2.4, were introduced independently in [43,46,60], and the duality theorems for crossed products, Theorems 2.5 and 2.7, were proved there. The proofs presented here are taken from [47] which resembles [60]. The key to the proof is in Lemma 2.10. Here we take an idea due to Van Heeswijck to [77]. On the other hand Landstad, [42], prepared Theorem II.2.1.(ii) in order to prove Theorem 2.7. Various results of this section were generalized to the Kac algebra context [22,26].

§3. <u>Roberts action and Tannaka-Tatsuuma duality</u>.

In this section, we shall discuss the duality for the "automorphism actions" of a locally compact group on a von Neumann algebra through a formalism given by Roberts.

In order to avoid unnecessary complications, we consider only compact groups in this section, while this restriction can be lifted without serious difficulties if one really needs to do so. We leave the general case to the reader.

<u>Definition</u> 3.1. A collection R of unitary representations of G is called a <u>ring</u> if i) $\pi_1 \oplus \pi_2 \in R$ and $\pi_1 \otimes \pi_2 \in R$ for every pair $\pi_1, \pi_2 \in R$; ii) The trivial representation ι of G belongs to R. If the conjugate representation $\bar{\pi}$ of each $\pi \in R$ falls in R again, then the ring R is said to be <u>self-adjoint</u>.

For each $\pi_1, \pi_2 \in R$, we denote

(3.1) $\qquad \mathcal{I}_G(\pi_2, \pi_1) = \{a \in \mathcal{L}(\mathfrak{H}_{\pi_1}, \mathfrak{H}_{\pi_2}) : a\pi_1(t) = \pi_2(t)a, \ t \in G\}$.

Let $\mathrm{End}(\mathfrak{n})$ be the set of all *-endomorphisms of $\mathfrak{n}$. Here we assume the normality and the identity preserving for endomorphisms. For each $\rho_1, \rho_2 \in \mathrm{End}(\mathfrak{n})$, we write

(3.2) $\qquad \mathcal{I}_\mathfrak{n}(\rho_2, \rho_1) = \{a \in \mathfrak{n} : a\rho_1(y) = \rho_2(y)a, \ y \in \mathfrak{n}\}$.

We then have the following relations among these sets:

(3.3)
$$
\begin{cases}
\mathcal{I}_G(\pi_3, \pi_2)\mathcal{I}_G(\pi_2, \pi_1) \subset \mathcal{I}_G(\pi_3, \pi_1) ; \\[4pt]
\mathcal{I}_G(\pi_2, \pi_1) \oplus \mathcal{I}_G(\pi_2', \pi_1') \subset \mathcal{I}_G(\pi_2 \oplus \pi_2', \pi_1 \oplus \pi_1') ; \\[4pt]
\mathcal{I}_G(\pi_2, \pi_1) \otimes \mathcal{I}_G(\pi_2', \pi_1') \subset \mathcal{I}_G(\pi_2 \otimes \pi_2', \pi_1 \otimes \pi_1') ; \\[4pt]
\mathcal{I}_\mathfrak{n}(\rho_3, \rho_2)\mathcal{I}_\mathfrak{n}(\rho_2, \rho_1) \subset \mathcal{I}_\mathfrak{n}(\rho_3, \rho_1) ; \\[4pt]
\rho(\mathcal{I}_\mathfrak{n}(\rho_2, \rho_1)) \subset \mathcal{I}_\mathfrak{n}(\rho \cdot \rho_2, \rho \cdot \rho_1) ; \\[4pt]
\mathcal{I}_\mathfrak{n}(\rho_2, \rho_1) \subset \mathcal{I}_\mathfrak{n}(\rho_2 \cdot \rho, \rho_1 \cdot \rho) ,
\end{cases}
$$
and

(3.4)
$$
\begin{cases}
\mathcal{I}_\mathfrak{n}(\rho_2, \rho_1)\rho_1(\mathcal{I}_\mathfrak{n}(\rho_2', \rho_1')) \subset \mathcal{I}_\mathfrak{n}(\rho_2 \cdot \rho_2', \rho_1 \cdot \rho_1') ; \\[4pt]
\rho_2'(\mathcal{I}_\mathfrak{n}(\rho_2, \rho_1))\mathcal{I}_\mathfrak{n}(\rho_2', \rho_1') \subset \mathcal{I}_\mathfrak{n}(\rho_2' \cdot \rho_2, \rho_1' \cdot \rho_1) .
\end{cases}
$$

<u>Definition</u> 3.2. A <u>Roberts</u> <u>action</u> $\{\rho, \eta\}$ of a ring R of representations of G on $\mathfrak{n}$ is a composition $\{\rho_\pi, \eta_{\pi_1, \pi_2} : \pi, \pi_1, \pi_2 \in R\}$, where $\rho_\pi \in \mathrm{End}(\mathfrak{n})$ and η_{π_1, π_2} is a σ-weakly continuous linear map of $\mathcal{I}_G(\pi_1, \pi_2)$ into $\mathcal{I}_\mathfrak{n}(\rho_{\pi_1}, \rho_{\pi_2})$ such that

i) $\rho_{\pi_1 \otimes \pi_2} = \rho_{\pi_1} \circ \rho_{\pi_2}$,

ii) $\eta_{\pi_2 \otimes \pi_2', \pi_1 \otimes \pi_1'}(a \otimes a') = \eta_{\pi_2, \pi_1}(a) \rho_{\pi_1}\left(\eta_{\pi_2', \pi_1'}(a')\right)$

$= \rho_{\pi_2}\left(\eta_{\pi_2', \pi_1'}(a')\right) \eta_{\pi_2, \pi_1}(a)$

for every $a \in \mathcal{I}_G(\pi_2, \pi_1)$ and $a' \in \mathcal{I}_G(\pi_2', \pi_1')$;

iii) $\eta_{\pi, \pi}(1) = 1$;

iv) $\eta_{\pi_1, \pi_2}(a)^* = \eta_{\pi_2, \pi_1}(a^*)$, $a \in \mathcal{I}_G(\pi_1, \pi_2)$;

v) $\eta_{\pi_1, \pi_2}(a) \eta_{\pi_2, \pi_3}(b) = \eta_{\pi_1, \pi_3}(ab)$

for every $a \in \mathcal{I}_G(\pi_1, \pi_2)$ and $b \in \mathcal{I}_G(\pi_2, \pi_3)$.

Before giving an important example of Roberts action, we need a few preparations concerning Hilbert spaces <u>in</u> a von Neumann algebra.

<u>Definition</u> 3.3. A Hilbert space <u>in</u> a von Neumann algebra $\mathfrak{h}$ is a closed subspace $\mathfrak{K}$ of $\mathfrak{h}$ with the following properties:

i) For every $x, y \in \mathfrak{K}$, y^*x is a scalar multiple of the identity; hence one can consider y^*x as the inner product $(x|y)$ of x and y;

ii) $a\mathfrak{K} \neq \{0\}$ whenever $a \neq 0$, $a \in \mathfrak{h}$.

It is easy to see that every element of $\mathfrak{K}$ with norm one is an isometry. Hence, if $\mathfrak{h}$ is finite, then every Hilbert space <u>in</u> $\mathfrak{h}$ is one dimensional and the scalar multiples of a unitary. So a Hilbert space in a finite von Neumann algebra is not interesting. Thus, we must consider properly infinite von Neumann algebras. A normalized orthogonal basis of a Hilbert space $\mathfrak{K}$ in $\mathfrak{h}$ is then a system $\{u_i : i = 1, \ldots, d\}$ of isometries with orthogonal ranges and $\sum_{i=1}^d u_i u_i^* = 1$. Once it is chosen, the map: $x \in \mathfrak{h} \to \sum_{i=1}^d u_i x u_i^*$ is an endomorphism of $\mathfrak{h}$ and does not depend on the choice of a basis; hence we denote it by $\rho_{\mathfrak{K}}$. One can characterize $\rho_{\mathfrak{K}}$ by the equality:

(3.5) $$\rho_{\mathfrak{K}}(a)x = xa , \quad x \in \mathfrak{K} , \quad a \in \mathfrak{h}$$

It is easy to check that for Hilbert spaces $\mathfrak{K}_1$ and $\mathfrak{K}_2$ in $\mathfrak{h}$, the σ-weak closure of the linear subspace spanned by xy^*, $x \in \mathfrak{K}_1$ and $y \in \mathfrak{K}_2$, is economically identified with $\mathfrak{L}(\mathfrak{R}_2,\mathfrak{R}_1)$; hence $\mathfrak{L}(\mathfrak{R}_2,\mathfrak{R}_1) \subset \mathfrak{n}$. Moreover, we have $\rho_{\mathfrak{R}}(\mathfrak{n}) = \mathfrak{L}(\mathfrak{R})' \cap \mathfrak{n}$; hence

$$\mathfrak{n} = \rho_{\mathfrak{R}}(\mathfrak{n}) \mathbin{\overline{\otimes}} \mathfrak{L}(\mathfrak{R}) .$$

An important feature of Hilbert spaces in $\mathfrak{n}$ is that for any pair $\mathfrak{R}_1,\mathfrak{R}_2$ of Hilbert spaces in $\mathfrak{n}$, the closed subspace spanned by the products xy, $x \in \mathfrak{R}_1$ and $y \in \mathfrak{R}_2$, is naturally identified with the tensor product $\mathfrak{R}_1 \otimes \mathfrak{R}_2$. Here the point is that the product $\mathfrak{R}_1\mathfrak{R}_2$ is a concrete object sitting in $\mathfrak{n}$ while $\mathfrak{R}_1 \otimes \mathfrak{R}_2$ is abstract. In the following situation, this point becomes clearer.

Let $\{\mathfrak{M},G,\alpha\}$ be a covariant system with $\mathfrak{M}^\alpha$ properly infinite. We denote by $\mathcal{H}_\alpha(\mathfrak{M})$ the collection of all Hilbert spaces in $\mathfrak{M}$ globally invariant under $\{\alpha_t : t \in G\}$. If $\mathfrak{R} \in \mathcal{H}_\alpha(\mathfrak{M})$, then we have, $x,y \in \mathfrak{R}$,

$$\begin{aligned}(\alpha_t(x)|\alpha_t(y)) &= \alpha_t(y)^*\alpha_t(x) = \alpha_t(y^*x)\\ &= \alpha_t((x|y)1) = (x|y) .\end{aligned}$$

Hence the restriction of α to $\mathfrak{R}$ is a unitary representation of G on $\mathfrak{R}$. We denote this representation by $\alpha_{\mathfrak{R}}$ or $\{\alpha,\mathfrak{R}\}$. Then $\mathcal{H}_\alpha(\mathfrak{M})$ turns out to be a collection of representations of G which is, in turn, a ring in the sense that

$$(3.6)\quad \begin{cases} \{\alpha,\mathfrak{H}_1\mathfrak{H}_2\} \cong \{\alpha,\mathfrak{H}_1\} \otimes \{\alpha,\mathfrak{H}_2\} \ ; \{\alpha,\mathbf{C}1\} \cong \{\iota,\mathfrak{H}_\iota\} \ ; \\ \{\alpha, w_1\mathfrak{H}_1 + w_2\mathfrak{H}_2\} \cong \{\alpha,\mathfrak{H}_1\} \oplus \{\alpha,\mathfrak{H}_2\} , \end{cases}$$

where w_1 and w_2 are isometries in $\mathfrak{M}^\alpha$ with $w_1w_1^* + w_2w_2^* = 1$.

It is not hard to see that $\rho_{\mathfrak{R}}$, $\mathfrak{R} \in \mathcal{H}_\alpha(\mathfrak{M})$, leaves $\mathfrak{M}^\alpha$ globally invariant, and also by (3.5) that

$$\mathcal{J}_G(\alpha_{\mathfrak{R}_2},\alpha_{\mathfrak{R}_1}) \subset \mathcal{J}_{\mathfrak{M}^\alpha}(\rho_{\mathfrak{R}_2},\rho_{\mathfrak{R}_1}) \subset \mathfrak{M}^\alpha .$$

We then set

$$(3.7)\quad \begin{cases} \rho_{\alpha_{\mathfrak{R}}}(x) = \rho_{\mathfrak{R}}(x) , \ x \in \mathfrak{M}^\alpha, \ \mathfrak{R} \in \mathcal{H}_\alpha(\mathfrak{M}) . \\ \eta_{\alpha_{\mathfrak{R}_2},\alpha_{\mathfrak{R}_1}}(a) = a , \ a \in \mathcal{J}_G(\alpha_{\mathfrak{R}_2},\alpha_{\mathfrak{R}_1}), \ \mathfrak{R}_1,\mathfrak{R}_2 \in \mathcal{H}_\alpha(\mathfrak{M}) . \end{cases}$$

A straightforward calculation shows that $\{\rho_{\alpha_{\mathfrak{R}}},\eta_{\alpha_{\mathfrak{R}_2},\alpha_{\mathfrak{R}_1}} : \mathfrak{R},\mathfrak{R}_1,\mathfrak{R}_2 \in \mathcal{H}_\alpha(\mathfrak{M})\}$ is indeed a Roberts action of $\mathcal{H}_\alpha(\mathfrak{M})$ on $\mathfrak{M}^\alpha$.

We now have the following Tannaka duality theorem in our context:

<u>Theorem</u> 3.4. Assume that G is compact. If every irreducible subrepresentation of $\{\alpha,\mathfrak{M}\}$ is equivalent to some representation in $\mathcal{H}_\alpha(\mathfrak{M})$, then each

$\sigma \in \mathrm{Aut}(\mathfrak{M}/\mathfrak{M}^{\alpha})$ leaving every member $\mathfrak{K} \in \mathcal{H}_{\alpha}(\mathfrak{M})$ globally invariant must be of the form α_r for some $r \in G$.[5)]

It should be pointed out that the above theorem can be generalized to a locally compact group, if we assume that $\mathcal{H}_{\alpha}(\mathfrak{M})$ contains a member equivalent to the regular representation of G. Thus, the Tatsuuma duality theorem in our context remains valid also.

Proof. For each $a,b \in \mathfrak{K}$, $\mathfrak{K} \in \mathcal{H}_{\alpha}(\mathfrak{M})$ we set

$$f_{a,b}(t) = a^*\alpha_t(b) ,$$

and denote by $C_{\alpha}(G)$ the set of all such functions. It is easy to check that $C_{\alpha}(G)$ is a $*$-subalgebra of $C(G)$, because $\mathcal{H}_{\alpha}(\mathfrak{M})$ is a self-adjoint ring.

Now, we define a map U of $C_{\alpha}(G)$ into itself by

$$Uf_{a,b} = f_{\sigma(a),b} ;$$

which is well defined because

$$\|Uf_{a,b}\|_2^2 = \int |f_{\sigma(a),b}(t)|^2 \, dt = \int |\sigma(a)^*\alpha_t(b)|^2 \, dt$$

$$= \sigma(a)^* \int \alpha_t(b\, b^*)dt \; \sigma(a)$$

$$= \sigma(a^* \int \alpha_t(b\, b^*) \, dt \; a)$$

$$= \sigma(\|f_{a,b}\|_2^2 \, 1) = \|f_{a,b}\|_2^2 .$$

It also follows that U is an isometry. Moreover, U is multiplicative. Indeed,

$$U(f_{a,b}f_{c,d}) = U f_{ac,bd} = f_{\sigma(ac),bd} = f_{\sigma(a),b}f_{\sigma(c),d} .$$

Since $U\rho_t = \rho_t U$ on $C_{\alpha}(G)$, Lemma 3.5 below tells us that $U = \lambda_r$ for some $r \in G$ on $C_{\alpha}(G)$. Therefore

$$\sigma(a)^*\alpha_s(b) = a^*\alpha_{r^{-1}s}(b) = \alpha_r(a)^*\alpha_s(b)$$

for all $a,b \in \mathfrak{K}$, $\mathfrak{K} \in \mathcal{H}_{\alpha}(\mathfrak{M})$. Therefore

$$\sigma(x) = \alpha_r(x)$$

for all $x \in \mathfrak{K}$, $\mathfrak{K} \in \mathcal{H}_{\alpha}(\mathfrak{M})$ and for all $x \in \mathfrak{M}^{\alpha}$.

5) $\mathrm{Aut}(\mathfrak{M}/\mathfrak{N})$ means the group of all automorphisms of $\mathfrak{M}$ leaving $\mathfrak{N}$ pointwise fixed.

If $\{\alpha,\mathfrak{M}\}$ is an irreducible subrepresentation of $\{\alpha,\mathfrak{m}\}$, then $\{\alpha,\mathfrak{M}\} \cong \{\alpha,\mathfrak{K}\}$ for some $\mathfrak{K} \in \mathfrak{H}_\alpha(\mathfrak{m})$ by assumption. Then there exists a basis $\{a_1,\dots,a_d\}$ of $\mathfrak{M}$ and an orthonormal basis $\{v_1,\dots,v_d\}$ of $\mathfrak{K}$ such that

$$\alpha_r(a_k) = \sum_j \lambda_{jk} a_j ; \quad \alpha_r(v_k) = \sum_j \lambda_{jk} v_j .$$

Put $w = \sum a_j v_j^*$. Then $w \in \mathfrak{m}^\alpha$ and $w\mathfrak{K} = \mathfrak{M}$. Therefore, if $a \in \mathfrak{K}$, then

$$\sigma(wa) = w\sigma(a) = w\alpha_r(a) = \alpha_r(wa) ,$$

so $\sigma = \alpha_r$ on $\mathfrak{M}$. Since G is compact, the collection of these spaces $\mathfrak{M}$ is total in $\mathfrak{m}$. Therefore, $\sigma = \alpha_r$ on $\mathfrak{m}$. Q.E.D.

Lemma 3.5. Let G be a compact group, A and B be a *-subalgebras of $C(G)$ globally invariant under ρ_t, $t \in G$. If σ is an isomorphism of A onto B such that

$$\text{i)} \quad \sigma \circ \rho_t = \rho_t \circ \sigma , \qquad t \in G ;$$
$$\text{ii)} \quad \|f\|_2 = \|\sigma f\|_2 , \qquad f \in A ,$$

then σ is of the form λ_r for some $r \in G$, hence preserves the adjoint operation in A.

Proof. Let $\mathfrak{M}$ and $\mathfrak{N}$ be the closures of A and B in $L^2(G)$ respectively. By assumption (ii), σ is extended to a unitary U of $\mathfrak{M}$ onto $\mathfrak{N}$. Let π be the multiplication representation of $C(G)$ on $L^2(G)$. We then have, for each $f \in A$, $g \in B$,

$$U\pi(f)U^{-1}g = \pi(\sigma f)g .$$

Thus, the unitary U gives rise to an isomorphism of the uniform closure of $\pi(A)|_{\mathfrak{M}}$ onto the uniform closure of $\pi(B)|_{\mathfrak{N}}$. However, the map: $f \in A \to \pi(f)|_{\mathfrak{M}} \in \mathcal{L}(\mathfrak{M})$ is extended to a faithful representation of the closure $[A]_\infty$ of A in $C(G)$, where the faithfulness of the extended representation follows from the fact that $[A]_\infty \subset \mathfrak{M}$. Furthermore, the extended representation is indeed given by restricting the space $\pi([A]_\infty)$ to $\mathfrak{M}$. Thus, the isomorphism of the closure of $\pi(A)|_{\mathfrak{M}}$, which is $\pi([A]_\infty))|_{\mathfrak{M}}$, onto the closure of $\pi(B)|_{\mathfrak{N}} (= \pi([B]_\infty)|_{\mathfrak{N}})$ gives rise to an isomorphism $\bar{\sigma}$ of $[A]_\infty$ onto $[B]_\infty$ which extends σ.

Noticing that every character of $[A]_\infty$ is given by evaluating each function in $[A]_\infty$ at some point in G, we can find an element $r' \in G$ such that $(\bar{\sigma}f)(e) = f(r')$ for every $f \in [A]_\infty$. We then have

$$(\sigma f)(t) = (\rho_t \cdot \sigma f)(e) = \rho_t f(r') = f(r't), \quad f \in A, \quad t \in G .$$

Thus, putting $r = r'^{-1}$, we get $\sigma = \lambda_r$ as required. Q.E.D.

NOTES

The materials presented in this section are mainly taken from [55]. See also [3].

§4. Supplementary formulas.

We shall give definitions of an action, a co-action and a crossed product with respect to $\mathcal{R}(G)'$:

An action α' of G on $\mathfrak{m}$ with respect to $\mathcal{R}(G)'$ is an isomorphism of $\mathfrak{m}$ into $\mathfrak{m} \bar{\otimes} L^\infty(G)$ satisfying

$$(4.1)\qquad (\alpha' \otimes \iota) \circ \alpha' = (\iota \otimes \alpha'_G) \circ \alpha' ,$$

where α'_G is an isomorphism of $L^\infty(G)$ into $L^\infty(G) \bar{\otimes} L^\infty(G)$ with $(\alpha'_G f)(s,t) = f(ts)$, $(\alpha'_G f = \mathrm{Ad}_{V_G'^*}(f \otimes 1))$. A co-action δ' of G on $\mathfrak{n}$ with respect to $\mathcal{R}(G)'$ is an isomorphism of $\mathfrak{n}$ into $\mathfrak{n} \bar{\otimes} \mathcal{R}(G)'$ satisfying

$$(4.2)\qquad (\delta' \otimes \iota) \circ \delta' = (\iota \otimes \delta'_G) \circ \delta' ,$$

where δ'_G is an isomorphism of $\mathcal{R}(G)'$ into $\mathcal{R}(G)' \bar{\otimes} \mathcal{R}(G)'$ with $\delta'_G \lambda(r) = \lambda(r) \otimes \lambda(r)$, $(\delta'_G(y) = \mathrm{Ad}_{W_G'}(y \otimes 1))$. The crossed products $\mathfrak{m} \times_{\alpha'} G$ and $\mathfrak{n} \times_{\delta'} G$ are defined as the von Neumann algebras $\alpha'(\mathfrak{m}) \vee (\mathbf{C} \otimes \mathcal{R}(G)')$ and $\delta'(\mathfrak{n}) \vee (\mathbf{C} \otimes L^\infty(G))$, respectively. The dual co-action $(\alpha')^\wedge$ and the dual action $(\delta')^\wedge$ are defined by

$$(4.3)\qquad (\alpha')^\wedge(y) = \mathrm{Ad}_{1 \otimes W_G'}(y \otimes 1) , \qquad y \in \mathfrak{m} \times_{\alpha'} G ;$$

$$(4.4)\qquad (\delta')^\wedge(x) = \mathrm{Ad}_{1 \otimes V_G'^*}(x \otimes 1) , \qquad x \in \mathfrak{n} \times_{\delta'} G .$$

Now, we shall list the associativity conditions for W_G, W_G', V_G and V_G':

$$(4.5)\qquad (W_G^* \otimes 1)(\iota \otimes \sigma)(W_G^* \otimes 1) = ((\iota \otimes \sigma)(W_G^* \otimes 1))(W_G^* \otimes 1) = (\iota \otimes \delta_G)(W_G^*) .$$

$$(4.6)\qquad (W_G \otimes 1)(1 \otimes \sigma)(W_G \otimes 1) = (\iota \otimes \delta_G)(W_G) .$$

$$(4.7)\qquad (W_G' \otimes 1)(\iota \otimes \sigma)(W_G' \otimes 1) = ((\iota \otimes \sigma)(W_G' \otimes 1))(W_G' \otimes 1) = (\iota \otimes \delta_G')(W_G') .$$

$$(4.8)\qquad (W_G'^* \otimes 1)(\iota \otimes \sigma)(W_G'^* \otimes 1) = (\iota \otimes \delta_G')(W_G'^*) .$$

$$(4.9)\qquad (V_G \otimes 1)(\iota \otimes \sigma)(V_G \otimes 1) = (\iota \otimes \alpha_G)(V_G) .$$

$$(4.10)\qquad (V_G' \otimes 1)(\iota \otimes \sigma)(V_G' \otimes 1) = (\iota \otimes \alpha_G)(V_G') .$$

$$(4.11)\qquad (V_G^* \otimes 1)(\iota \otimes \sigma)(V_G^* \otimes 1) = (\iota \otimes \alpha_G')(V_G^*) .$$

$$(4.12)\qquad (V_G'^* \otimes 1)(\iota \otimes \sigma)(V_G'^* \otimes 1) = (\iota \otimes \alpha_G')(V_G'^*) .$$

CHAPTER II.

ELEMENTARY PROPERTIES OF CROSSED PRODUCTS.

Introduction. The image of the original algebra in the crossed product is characterized as the fixed point subalgebra under the dual action (resp. the dual co-action), Theorem 1.1, in §1. Combining this with the duality theorem, the crossed product is characterized as the fixed point subalgebra of the tensor product of the original algebra with $\mathcal{L}(L^2(G))$ under the tensor product of the original (resp. co-)action and the regular (resp.co-)action, Theorem 1.2.

Section 2 is devoted to a characterization of a dual(resp. co-)action. The characterization should be viewed as a sort of an imprimitivity theorem.

In §3, we consider the commutant of the crossed product by a closed subgroup which is shown to be the fixed point subalgebra of the crossed product of the commutant of the original algebra under the action (or co-action) of the quotient group. Here, since we do not treat the Banach algebra bundle of Fell, [30], we have to restrict ourselves to normal subgroups for actions.

§1. Fixed points in crossed products.

Given an action α of G on $\mathfrak{M}$ or a co-action δ of G on $\mathfrak{N}$, we denote by $\mathfrak{M}^\alpha$ or $\mathfrak{N}^\delta$ the fixed point subalgebra of $\mathfrak{M}$ or $\mathfrak{N}$, i.e.,

$$\mathfrak{M}^\alpha = \{x \in \mathfrak{M} : \alpha(x) = x \otimes 1\} ;$$
$$\mathfrak{N}^\delta = \{x \in \mathfrak{N} : \delta(x) = x \otimes 1\} .$$

With these notations, we can characterize the location of the original von Neumann algebra (or more precisely the image under the action or the co-action) in the crossed product as follows:

Theorem 1.1. (a) $(\mathfrak{M} \times_\alpha G)^{\hat{\alpha}} = \alpha(\mathfrak{M})$.
(b) $(\mathfrak{N} \times_\delta G)^{\hat{\delta}} = \delta(\mathfrak{N})$.

Proof. (a) It is clear that $\alpha(\mathfrak{M})$ is contained in $(\mathfrak{M} \times_\alpha G)^{\hat{\alpha}}$. We have only to show the reverse inclusion.

Since $\{\mathfrak{M},\alpha\} \cong \{\alpha(\mathfrak{M}), \iota \otimes \rho\}$ and $\iota \otimes \rho_t = \mathrm{Ad}_{1\otimes\rho(t)}$, we may assume that the action α is implemented by a unitary representation u of G such that $\alpha_t(x) = u(t)xu(t)^*$ for $x \in \mathfrak{M}$. Here we identify u with the unitary in $\mathcal{L}(\mathfrak{H}) \bar{\otimes} L^\infty(G)$ given by $(u\xi)(t) = u(t)\xi(t)$ for $\xi \in \mathfrak{H} \otimes L^2(G)$. Then $\alpha(\mathfrak{M}) = u(\mathfrak{M} \otimes \mathbf{C})u^*$. We put

$$\alpha'(x) = u^*(x \otimes 1)u , \quad x \in \mathfrak{M}' .$$

Then α' is an action of G on $\mathfrak{M}'$ with respect to $\mathcal{R}'(G)$. By Lemma I.2.6, $\mathfrak{M}' \bar{\otimes} L^\infty(G)$ is generated by $\alpha'(\mathfrak{M}')$ and $\mathbf{C} \otimes L^\infty(G)$. Therefore

$$\mathfrak{M}' \bar{\otimes} \mathcal{L}(L^2(G)) = \alpha'(\mathfrak{M}') \vee (\mathbf{C} \otimes L^\infty(G)) \vee (\mathbf{C} \otimes \mathcal{R}'(G)) .$$

Applying Ad_u to the both sides and considering the commutants, we have

$$(1.1) \qquad \alpha(\mathfrak{M}) = (\mathfrak{M}' \otimes \mathbf{C})' \cap (\mathbf{C} \otimes L^\infty(G))' \cap \{u(t) \otimes \lambda(t) : t \in G\}' .$$

Now suppose that $y \in \mathfrak{M} \times_\alpha G$ and $\hat{\alpha}(y) = y \otimes 1$. We want to show y belongs to the right hand side of (1.1). It is straight forward to see that $\mathfrak{M} \times_\alpha G$ commutes with $\mathfrak{M}' \otimes \mathbf{C}$ and $u(t) \otimes \lambda(t)$ for all $t \in G$. Thus y commutes with $\mathfrak{M}' \otimes \mathbf{C}$ and $u(t) \otimes \lambda(t)$ for all $t \in G$. Since $\hat{\alpha}(y) = (1 \otimes W_G^*)(y \otimes 1)(1 \otimes W_G)$, $y \otimes 1$ commutes with

$$\mathbf{C} \otimes \mathbf{C} \otimes L^\infty(G) \quad \text{and} \quad (1 \otimes W_G^*)(\mathbf{C} \otimes \mathbf{C} \otimes L^\infty(G))(1 \otimes W_G) .$$

Since, by Lemma I.2.6

$$(\mathbf{C} \otimes L^\infty(G)) \vee W_G^*(\mathbf{C} \otimes L^\infty(G))W_G = L^\infty(G) \bar{\otimes} L^\infty(G) ,$$

it follows that y commutes with $\mathbf{C} \otimes L^\infty(G)$.

(b) We have only to show that $(\mathcal{M} \times_\delta G)^{\hat{\hat{\delta}}}$ is contained in $\delta(\mathcal{M})$. Since $\{\mathcal{M},\delta\} \cong \{\delta(\mathcal{M}), \iota \otimes \delta_G\}$ and $\delta_G(x) = W_G^*(x \otimes 1)W_G$, we may assume that δ is implemented by a unitary w in $\mathcal{L}(\mathfrak{K}) \bar{\otimes} \mathcal{R}(G)$ such that

$$(1.2) \qquad \begin{aligned} &\delta(y) = w^*(y \otimes 1)w\,, \qquad y \in \mathcal{M}\,, \\ &(w \otimes 1)(\iota \otimes \sigma)(w \otimes 1) = (\iota \otimes \delta_G)(w)\,. \end{aligned}$$

Let K be a unitary on $L^2(G)$ defined by $(K\xi)(t) = \Delta(t)^{1/2}\xi(t^{-1})$ for $\xi \in L^2(G)$ and $w' = (1 \otimes K)w(1 \otimes K)$. Since $w^*(\mathcal{M} \otimes \mathbf{C})w \subset \mathcal{M} \bar{\otimes} \mathcal{L}(L^2(G))$ and $w \in \mathcal{L}(\mathfrak{K}) \bar{\otimes} \mathcal{R}(G)$, it follows that

$$w(\mathcal{M}' \otimes \mathbf{C})w^* \subset \mathcal{M}' \bar{\otimes} \mathcal{R}(G)\,.$$

Here we set

$$\delta'(x) = w'(x \otimes 1)w'^*, \qquad x \in \mathcal{M}'\,.$$

Then δ' is an isomorphism of $\mathcal{M}'$ into $\mathcal{M}' \bar{\otimes} \mathcal{R}(G)'$ and w' satisfies $w' \in \mathcal{L}(\mathfrak{K}) \bar{\otimes} \mathcal{R}'(G)$ and

$$(w' \otimes 1)(\iota \otimes \sigma)(w' \otimes 1) = (\iota \otimes \delta_G')(w')\,.$$

Therefore, δ' is a co-action of G on $\mathcal{M}'$ with respect to $\mathcal{R}'(G)$. By Lemma I.2.10, $\mathcal{M}' \bar{\otimes} \mathcal{L}(L^2(G))$ is generated by $\delta'(\mathcal{M}')$ and $\mathbf{C} \otimes \mathcal{L}(L^2(G))$. Therefore

$$\mathcal{M}' \bar{\otimes} \mathcal{L}(L^2(G)) = \delta'(\mathcal{M}') \vee (\mathbf{C} \otimes L^\infty(G)) \vee (\mathbf{C} \otimes \mathcal{R}(G))\,.$$

Applying $\mathrm{Ad}_{w^*(1 \otimes K)}$ to the both sides and considering the commutants, we have

$$\delta(\mathcal{M}) = (\mathcal{M}' \otimes \mathbf{C})' \cap (w^*(\mathbf{C} \otimes L^\infty(G))w)' \cap (\mathbf{C} \otimes \mathcal{R}'(G))'\,.$$

Suppose that $y \in \mathcal{M} \times_\delta G$ and $\hat{\delta}(y) = y \otimes 1$. It is clear that $\mathcal{M} \times_\delta G$ commutes with $\mathcal{M}' \otimes \mathbf{C}$. Since (1.2) implies

$$(1 \otimes W_G)(w^* \otimes 1)(\iota \otimes \sigma)(w^* \otimes 1) = (w^* \otimes 1)(1 \otimes W_G)\,,$$

we have

$$\mathrm{Ad}_{1 \otimes W_G}((w^*(1 \otimes f)w) \otimes 1) = (w^*(1 \otimes f)w) \otimes 1$$

for $f \in L^\infty(G)$. It is known that $W_G(x \otimes 1)W_G^* = x \otimes 1$ if and only if $x \in L^\infty(G)$. Thus $w^*(1 \otimes f)w$ commutes with $\mathbf{C} \otimes L^\infty(G)$. Therefore, $\mathcal{M} \times_\delta G$ commutes with $w^*(\mathbf{C} \otimes L^\infty(G))w$. Thus y commutes with $\mathcal{M}' \otimes \mathbf{C}$ and $w^*(\mathbf{C} \otimes L^\infty(G))w$. Since $\hat{\delta}(y) = (1 \otimes V_G')(y \otimes 1)(1 \otimes V_G')^*$, $y \otimes 1$ commutes with

$$\mathbf{C} \otimes \mathbf{C} \otimes \mathcal{R}'(G) \qquad \text{and} \qquad (1 \otimes V_G')(\mathbf{C} \otimes \mathbf{C} \otimes \mathcal{R}'(G))(1 \otimes V_G')^*\,.$$

Therefore, y commutes with $\mathbf{C} \otimes \mathcal{R}'(G)$. Q.E.D.

Combining Theorem 1.1 and the duality theorem for crossed products, we have a characterization of crossed products.

Theorem 1.2. (a) $\mathfrak{M} \times_\alpha G = (\mathfrak{M} \overline{\otimes} \mathcal{L}(L^2(G)))^{\tilde{\alpha}}$.

(b) $\mathfrak{N} \times_\delta G = (\mathfrak{N} \overline{\otimes} \mathcal{L}(L^2(G)))^{\tilde{\delta}}$.

Proof. (a) Making use of the isomorphism π of $\mathfrak{M} \overline{\otimes} \mathcal{L}(L^2(G))$ onto $(\mathfrak{M} \times_\alpha G) \times_{\hat{\alpha}} G$ obtained in (I.2.12), we have

$$\pi((\mathfrak{M} \overline{\otimes} \mathcal{L}(L^2(G)))^{\tilde{\alpha}}) = ((\mathfrak{M} \times_\alpha G) \times_{\hat{\alpha}} G)^{\hat{\hat{\alpha}}}$$

by the duality theorem for action. The right hand side is $\hat{\alpha}(\mathfrak{M} \times_\alpha G)$ by Theorem 1.1. On the other hand, $\hat{\alpha} = \pi$ on $\mathfrak{M} \times_\alpha G$ by (I.2.8), (I.2.9) and (I.2.13).

(b) Making use of the isomorphism π of $\mathfrak{N} \overline{\otimes} \mathcal{L}(L^2(G))$ onto $(\mathfrak{N} \times_\delta G) \times_{\hat{\delta}} G$ obtained in (I.2.20), we have

$$\pi((\mathfrak{N} \overline{\otimes} \mathcal{L}(L^2(G)))^{\tilde{\delta}}) = ((\mathfrak{N} \times_\delta G) \times_{\hat{\delta}} G)^{\hat{\hat{\delta}}}$$

by the duality theorem for co-action. The right hand side is $\hat{\delta}(\mathfrak{N} \times_\delta G)$ by Theorem 1.1. On the other hand, $\hat{\delta} = \pi$ on $\mathfrak{N} \times_\delta G$ by (I.2.10), (I.2.11), (I.2.21) and (I.2.22). Q.E.D.

NOTES

The fixed points in the crossed product, Theorem 1.1, are characterized in [42,43,46,60]. The presented proof is taken from [46]. The characterizations of the crossed product as the fixed point subalgebra, Theorem 1.2, were obtained by Digerness [15,16] and by [43,47,60]. Theorems 1.1 and 1.2 are generalized to the Kac algebra context in [26].

§2. Characterization of crossed products.

When one studies an action or a co-action of a locally compact group G on a von Neumann algebra, often it so happens that the action or the co-action is already dual to another co-action or action. Thus, we need to know exactly when the given one is dual to something else. The following gives a convenient characterization for those actions and co-actions.

Theorem 2.1. Let δ be a co-action of G on $\mathfrak{n}$. The following three conditions are equivalent:

i) There exist a von Neumann algebra $\mathfrak{m}$ and an action α of G on $\mathfrak{m}$ such that

$$\{\mathfrak{n},\delta\} \cong \{\mathfrak{m} \times_\alpha G, \hat{\alpha}\} \ ;$$

ii) There exists a unitary representation u of G in $\mathfrak{n}$ such that

$$\delta(u(t)) = u(t) \otimes \rho(t) \ , \qquad t \in G \ ; \tag{2.1}$$

iii) There exists a unitary u in $\mathfrak{n} \,\overline{\otimes}\, L^\infty(G)$ such that

$$(u \otimes 1)(\iota \otimes \sigma)(u \otimes 1) = (\iota \otimes \alpha_G)(u) \ ; \tag{2.2}$$

$$\bar{\delta}(u) = (u \otimes 1)(1 \otimes W_G) \ , \tag{2.3}$$

where $\bar{\delta} = (\iota \otimes \sigma) \circ (\delta \otimes \iota)$.

If any of these three conditions holds, then $\mathfrak{n}$ is generated by $\mathfrak{n}^\delta$ and $u(t)$, $t \in G$.

Proof. (i) $\Rightarrow$ (ii): Trivial.

(ii) $\Rightarrow$ (iii): Given a unitary representation u of G in $\mathfrak{n}$ satisfying (2.1), we shall use the same symbol u for the unitary in $\mathfrak{n} \,\overline{\otimes}\, L^\infty(G)$ defined by $(u\xi)(t) = u(t)\xi(t)$ for $\xi \in \mathfrak{K} \otimes L^2(G)$. Then $u(st) = u(s)u(t)$ means (2.2). Furthermore, (2.1) means precisely (2.3). Namely, if $\xi \in \mathfrak{K} \otimes L^2(G \times G)$ and $U_\sigma\xi(s,t) = \xi(t,s)$, then

$$\begin{aligned}\bar{\delta}(u)\xi(s,t) &= U_\sigma(\delta \otimes \iota)(u)(U_\sigma\xi)(s,t) = (\delta \otimes \iota)(u)(U_\sigma\xi)(t,s) \\ &= \delta(u(s))(U_\sigma\xi)(t,s) = (u(s) \otimes \rho(s))U_\sigma\xi(t,s) \\ &= u(s)(U_\sigma\xi)(ts,s) = u(s)\xi(s,ts) \\ &= (u \otimes 1)(1 \otimes W_G)\xi(s,t) \ .\end{aligned}$$

(iii) $\Rightarrow$ (i): For each $x \in \mathfrak{n}^\delta$, we set $\alpha(x) = u(x \otimes 1)u^*$. Then α is an isomorphism of $\mathfrak{n}^\delta$ into $\mathfrak{n} \,\overline{\otimes}\, L^\infty(G)$. By (2.3), we have, for each $x \in \mathfrak{n}^\delta$,

$$\bar{\delta} \circ \alpha(x) = \bar{\delta}(u(x \otimes 1)u^*) = \bar{\delta}(u)\bar{\delta}(x \otimes 1)\bar{\delta}(u)^*$$
$$= (u \otimes 1)(1 \otimes W_G)[(\iota \otimes \sigma) \circ (\delta \otimes \iota)(x \otimes 1)](1 \otimes W_G^*)(u^* \otimes 1)$$
$$= (u \otimes 1)(1 \otimes W_G)[(\iota \otimes \sigma)(\delta(x) \otimes 1)](1 \otimes W_G^*)(u^* \otimes 1)$$
$$= (u \otimes 1)(1 \otimes W_G)[(\iota \otimes \sigma)(x \otimes 1_{L^2(G\times G)})](1 \otimes W_G^*)(u^* \otimes 1)$$
$$= (u \otimes 1)(x \otimes 1)(u^* \otimes 1) = \alpha(x) \otimes 1 .$$

Therefore, $\alpha(x)$ falls in $\mathfrak{n}^\delta \bar{\otimes} L^\infty(G)$, so that α maps $\mathfrak{n}^\delta$ into $\mathfrak{n}^\delta \bar{\otimes} L^\infty(G)$. By (2.2), α satisfies the associativity condition, (I.2.4); hence it is an action of G on $\mathfrak{n}^\delta$.

Next, we want to show that $\mathfrak{n}$ is generated by $\mathfrak{n}^\delta$ and $u(G)$. By Lemma I.2.10, the von Neumann algebra $\bar{\mathfrak{n}} = \mathfrak{n} \bar{\otimes} \mathfrak{L}(L^2(G))$ is generated by $\mathfrak{n} \times_\delta G$ and $\mathbf{C} \otimes \mathcal{R}(G)'$. By Theorem 1.2, $\mathfrak{n} \times_\delta G = \bar{\mathfrak{n}}^{\bar{\delta}}$. Since Ad_u is an isomorphism of $\{\bar{\mathfrak{n}},\tilde{\delta}\}$ onto $\{\bar{\mathfrak{n}},\bar{\delta}\}$ by (2.3), it follows that $\bar{\mathfrak{n}}$ is also generated by $\bar{\mathfrak{n}}^{\tilde{\delta}} = \mathfrak{n}^\delta \bar{\otimes} \mathfrak{L}(L^2(G))$ and $u(\mathbf{C} \otimes \mathcal{R}(G)')u^*$. Since $u(1 \otimes \lambda(r))u^* = u(r) \otimes \lambda(r)$, $\mathfrak{n}$ must be generated by $\mathfrak{n}^\delta$ and $u(G)$.

Finally, we shall show that $\{\mathfrak{n},\delta\} \cong \{\mathfrak{n}^\delta \times_\alpha G,\hat{\alpha}\}$. For each $x \in \mathfrak{n}^\delta$, we have

$$\text{(2.4)} \qquad \begin{cases} u\delta(x)u^* = u(x \otimes 1)u^* = \alpha(x) \\ u\delta(u(t))u^* = u(u(t) \otimes \rho(t))u^* = 1 \otimes \rho(t) . \end{cases}$$

Therefore, $\mathrm{Ad}_u \circ \delta$ is an isomorphism $\mathfrak{n} = \mathfrak{n}^\delta \vee u(G)''$ onto $\mathfrak{n}^\delta \times_\alpha G$. By (2.4), we have

$$\mathrm{Ad}_{u\otimes 1} \circ (\delta \otimes \iota) \circ \delta = \hat{\alpha} \circ \mathrm{Ad}_u \circ \delta$$

on $\mathfrak{n}^\delta$ and $u(G)''$, which yields our assertion. Q.E.D.

The dual version of the above theorem is the following:

<u>Theorem</u> 2.2. Let α be an action of G on $\mathfrak{m}$. The following three conditions are equivalent:

i) There exist a von Neumann algebra $\mathfrak{n}$ and a co-action δ of G on $\mathfrak{n}$ such that

$$\{\mathfrak{m},\alpha\} \cong \{\mathfrak{n} \times_\delta G,\hat{\delta}\} .$$

ii) There exists a $*$-isomorphism π of $L^\infty(G)$ into $\mathfrak{m}$ such that

$$\text{(2.5)} \qquad \alpha_t \circ \pi = \pi \circ \lambda_t \quad (\text{or } \alpha \circ \pi = (\pi \otimes \iota) \circ \lambda) \quad \text{on } L^\infty(G) .$$

iii) There exists a unitary w in $\mathfrak{m} \bar{\otimes} \mathcal{R}(G)$ such that

$$\text{(2.6)} \qquad (w^* \otimes 1)(\iota \otimes \sigma)(w^* \otimes 1) = (\iota \otimes \delta_G)(w^*)$$

$$\text{(2.7)} \qquad \bar{\alpha}(w^*) = (w^* \otimes 1)(1 \otimes V_G) \quad (\text{or } (\alpha_t \otimes 1)(w^*) = w^*(1 \otimes \rho(t))) .$$

In the case (ii), $\mathfrak{m}$ is generated by $\mathfrak{m}^\alpha$ and $\pi(L^\infty(G))$.

Proof. (i) $\Rightarrow$ (ii): Clear.

(ii) $\Rightarrow$ (iii): We may assume that $\mathfrak{M}$ is standard and α is implemented by a unitary representation u of G on $\mathfrak{H}$. Then (2.5) implies $\mathrm{Ad}_{u(t)} \circ \pi = \pi \circ \lambda_t$ on $L^\infty(G)$. By Mackey's imprimitivity theorem, there exist a Hilbert space $\mathfrak{K}$ and an isometry U of $\mathfrak{K} \otimes L^2(G)$ onto $\mathfrak{H}$ such that

$$\pi(f) = U(1 \otimes f)U^{-1} \quad , \quad u(t) = U(1 \otimes \lambda(t))U^{-1} .$$

Therefore $U(\mathbb{C} \otimes L^\infty(G))U^{-1} \subset \mathfrak{M}$ and $\alpha_t \circ \mathrm{Ad}_U = \mathrm{Ad}_U \circ (\iota \otimes \lambda_t)$ on $\mathbb{C} \otimes L^\infty(G)$. Therefore we may assume that

$$\mathfrak{H} = \mathfrak{K} \otimes L^2(G) \quad , \quad \pi(f) = 1_{\mathfrak{K}} \otimes f \quad , \quad u(t) = v(t) \otimes \lambda(t)$$

for some unitary representation v of G on $\mathfrak{K}$, for $(1 \otimes \lambda(t))^* u(t) \in \mathcal{L}(\mathfrak{K}) \bar{\otimes} L^\infty(G)$.

Now, we define a unitary w in $\mathcal{L}(\mathfrak{H}) \bar{\otimes} \mathcal{R}(G)$ by $1_{\mathfrak{K}} \otimes W_G$. Since W_G^* satisfies the associativity condition (I.4.5), w^* satisfies (2.6). Since $1_{\mathfrak{K}} \otimes W_G \in \mathbb{C}_{\mathfrak{K}} \otimes L^\infty(G) \bar{\otimes} \mathcal{R}(G)$ and $\pi(L^\infty(G)) \subset \mathfrak{M}$, it follows that $w \in \mathfrak{M} \bar{\otimes} \mathcal{R}(G)$. Since

$$\mathrm{Ad}_{(\iota \otimes \sigma)(V_G' \otimes 1)}(W_G^* \otimes 1) = (W_G^* \otimes 1)(1 \otimes V_G) ,$$

the condition (2.7) is obtained.

(iii) $\Rightarrow$ (ii) Put $w_K = (1 \otimes K)w$ and $\pi_0(f) = w_K(1 \otimes f)w_K^*$ for $f \in L^\infty(G)$, where $(K\xi)(t) = \Delta(t)^{1/2}\xi(t^{-1})$ for $\xi \in L^2(G)$. Then $\pi_0(f) \in \mathfrak{M} \bar{\otimes} \mathcal{L}(L^2(G))$. Since (2.7) implies

$$\bar{\alpha}(w_K^*) = (w_K^* \otimes 1)(1 \otimes V_G') , \tag{2.8}$$

it follows that $\mathrm{Ad}_{w_K \otimes 1} \circ \bar{\alpha} = \tilde{\alpha} \circ \mathrm{Ad}_{w_K}$. Since $\bar{\mathfrak{M}}^{\tilde{\alpha}} = \mathfrak{M} \times_\alpha G$ by Theorem 1.2, $\pi_0(f) \in \mathfrak{M} \times_\alpha G$.

The associativity condition (2.6) implies that

$$(1 \otimes W_G)(w \otimes 1)(\iota \otimes \sigma)(w \otimes 1) = (w \otimes 1)(1 \otimes W_G) .$$

Since $W_G^* = (K \otimes 1)W_G(K \otimes 1)$, we have

$$(1 \otimes W_G^*)(w_K \otimes 1)(\iota \otimes \sigma)(w \otimes 1) = (w_K \otimes 1)(1 \otimes W_G) ,$$

and hence

$$\mathrm{Ad}_{1 \otimes W_G^*}(\pi_0(f) \otimes 1) = \pi_0(f) \otimes 1 .$$

Since $\alpha(\mathfrak{M}) = (\mathfrak{M} \times_\alpha G)^{\hat{\alpha}}$ by Theorem 1.1, we have $\pi_0(f) \in \alpha(\mathfrak{M})$.

Therefore there exists an isomorphism π of $L^\infty(G)$ into $\mathfrak{M}$ such that $\pi_0 = \alpha \circ \pi$. Since $(\alpha \otimes \iota) \circ \alpha = (\iota \otimes \alpha_G) \circ \alpha$, it follows that $\alpha \circ \alpha_t = (\iota \otimes \rho_t) \circ \alpha$ and hence that

$$\alpha \circ \alpha_t \circ \pi = (\iota \otimes \rho_t) \circ \alpha \circ \pi = (\iota \otimes \rho_t) \circ \pi_o$$
$$= \pi_o \circ \lambda_t = \alpha \circ \pi \circ \lambda_t$$

on $L^\infty(G)$. Thus $\alpha_t \circ \pi = \pi \circ \lambda_t$ on $L^\infty(G)$.

(ii) and (iii) $\Rightarrow$ (i): We set the map δ on $\mathfrak{M}^\alpha$ by

$$\delta(y) = w^*(y \otimes 1)w .$$

It follows from (2.7) that

$$\begin{aligned}\bar{\alpha} \circ \delta(y) &= \bar{\alpha} \circ \mathrm{Ad}_{w^*}(y \otimes 1)\\ &= \mathrm{Ad}_{w^*\otimes 1} \circ \mathrm{Ad}_{1\otimes v_G} \circ \bar{\alpha}(y \otimes 1) \qquad ,\\ &= \delta(y) \otimes 1 .\end{aligned}$$

Therefore $\delta(y) \in \mathfrak{M}^\alpha \bar{\otimes} \mathcal{L}(L^2(G))$. Since $w \in \mathfrak{M} \bar{\otimes} \mathcal{R}(G)$, δ is an isomorphism of $\mathfrak{M}^\alpha$ into $\mathfrak{M}^\alpha \bar{\otimes} \mathcal{R}(G)$. Moreover, w satisfies the associativity condition (2.6) and hence δ is a co-action of G on $\mathfrak{M}^\alpha$.

By Lemma I.2.6, $\bar{\mathfrak{M}}$ is generated by $\mathfrak{M} \times_\alpha G$ and $\mathbb{C} \otimes L^\infty(G)$. By Theorem 1.2, $\bar{\mathfrak{M}}^{\tilde{\alpha}} = \mathfrak{M} \times_\alpha G$. Condition (2.8) gives an isomorphism $\mathrm{Ad}_{w_K^*}$ of $\{\bar{\mathfrak{M}},\tilde{\alpha}\}$ onto $\{\bar{\mathfrak{M}},\bar{\alpha}\}$. Therefore $\bar{\mathfrak{M}}$ is generated by $\bar{\mathfrak{M}}^{\bar{\alpha}}$ and $w_K^*(\mathbb{C} \otimes L^\infty(G))w_K$.

According to (ii) there exist a Hilbert space $\mathfrak{K}$ and an isomorphism π of $L^\infty(G)$ onto $\mathfrak{M}$ such that

$$\mathfrak{H} = \mathfrak{K} \otimes L^2(G) \ , \quad \pi(f) = 1_{\mathfrak{K}} \otimes f \ , \quad \alpha \circ \pi = (\pi \otimes \iota) \circ \lambda .$$

Since $w_K^*(\mathbb{C}_{\mathfrak{H}} \otimes L^\infty(G))w_K \subset \mathbb{C}_{\mathfrak{K}} \otimes L^\infty(G) \bar{\otimes} L^\infty(G)$, $\mathfrak{M}$ is generated by $\mathfrak{M}^\alpha$ and $\pi(L^\infty(G))$.

Finally we shall show that $\{\mathfrak{M},\alpha\} \cong \{\mathfrak{M}^\alpha \times_\delta G,\hat{\delta}\}$. If $y \in \mathfrak{M}^\alpha$ and $f \in L^\infty(G)$, then

$$w_K^*\alpha(y)w_K = w_K^*(y \otimes 1)w_K = \delta(y) \tag{2.9}$$

$$w_K^*\alpha(\pi(f))w_K = w_K^*\pi_0(f)w_K = 1_{\mathfrak{H}} \otimes f \qquad (\alpha \circ \pi = \pi_0) . \tag{2.10}$$

Therefore $\mathrm{Ad}_{w_K^*} \circ \alpha$ is an isomorphism of $\mathfrak{M}$ onto $\mathfrak{M}^\alpha \times_\delta G$. Since, by (2.9) and (2.10),

$$\mathrm{Ad}_{w_K^*\otimes 1} \circ (\alpha \otimes \iota) \circ \alpha = \hat{\delta} \circ \mathrm{Ad}_{w_K^*} \circ \alpha$$

on $\mathfrak{M}^\alpha$ and $\pi(L^\infty(G))$. Q.E.D.

Here we generalize the concept of the dual action and the dual co-action defined in Proposition I. 2.4 as follows:

Definition 2.3. (a) An action α of G on $\mathfrak{m}$ is said to be dual or the dual action of G on $\mathfrak{m}$, if $\{\mathfrak{m},\alpha\} \cong \{\mathfrak{n} \times_\delta G, \hat{\delta}\}$ for some $\{\mathfrak{n},\delta\}$.

(b) A co-action δ of G on $\mathfrak{n}$ is said to be dual or the dual co-action of G on $\mathfrak{n}$, if $\{\mathfrak{n},\delta\} \cong \{\mathfrak{m} \times_\alpha G, \hat{\alpha}\}$ for some $\{\mathfrak{m},\alpha\}$.

In the rest of this section, we assume that G is discrete. Then there exists a bijection between the set of all unitaries w in $\mathfrak{m} \bar{\otimes} \mathcal{R}(G)$ satisfying the associativity condition

$$(w \otimes 1)(\iota \otimes \sigma)(w \otimes 1) = (\iota \otimes \delta_G)(w)$$

and the set of all partitions $\{e(t) : t \in G\}$ of the identity in $\mathfrak{m}$ by the relation

$$w = \sum_{t\in G} e(t) \otimes \rho(t) . \tag{2.11}$$

Thus our Theorems 2.1 and 2.2 are stated in terms of these partitions.

Theorem 2.4. (a) If δ is a co-action of G on $\mathfrak{n}$ implemented by a unitary $w \in \mathfrak{n} \bar{\otimes} \mathcal{R}(G)$ satisfying the associativity condition so that $\delta(y) = \mathrm{Ad}_{w^*}(y \otimes 1)$, then the following two conditions are equivalent:

(i) δ is dual;

(ii) there exists a unitary representation u of G in $\mathfrak{n}$ such that $u(t)e(r) = e(rt^{-1})u(t)$ for all r,t.

(b) If α is an action of G on $\mathfrak{m}$, then the following two conditions are equivalent:

(i) α is dual;

(ii) there exists a strictly wandering projection $e \in \mathfrak{m}$ for α, i.e. $\{\alpha_t(e) : t \in G\}$ is a partition of the identity such that $\alpha_t(e)\alpha_s(e) = 0$ for $t \neq s$.

Proof. (a) Condition (2.1) is equivalent to

$$\sum_r u(t)e(r) \otimes \rho(r) = \sum_s e(s)u(t) \otimes \rho(st) ,$$

which is equivalent to $u(t)e(r) = e(rt^{-1})u(t)$ for all t,r.

(b) Condition (2.6) is equivalent to the existence of a partition $\{e(t) : t \in G\}$ with (2.11). Condition (2.7) is equivalent to

$$\sum_r \alpha_t(e(r)) \otimes \rho(r^{-1}) = \sum_s e(s) \otimes \rho(s^{-1}t) .$$

which is equivalent to $\alpha_t(e(r)) = e(tr)$. Thus we have only to set $e = e(r)$ for some r. Q.E.D.

NOTES

The equivalence of (i) and (ii) in Theorem 2.1 is due to Landstad [42]; that of (i) and (ii) in Theorem 2.2 is due to Landstad [43], Nakagami [46] and Strătilă - Voiculescu - Zsidó [60]. Theorems 2.1 and 2.2 are generalized to the Kac algebra context in [26].

§3. Commutants of crossed products.

In this section we shall define an action β of G on the commutant of $\alpha(\mathfrak{M})$ and a co-action ε of G on the commutant of $\delta(\mathfrak{N})$ in order to show an imprimitivity theorem, which will be applied to the commutants of crossed products.

The associativity (I.2.4) for α and (I.2.5) for δ gives us

$$\mathrm{Ad}_{(1\otimes V_G)}(\alpha(\mathfrak{M}) \otimes \mathbf{C}) \subset (\alpha \otimes \iota)(\mathfrak{M} \,\bar{\otimes}\, L^\infty(G)) \ ;$$

$$\mathrm{Ad}_{(1\otimes W_G^*)}(\delta(\mathfrak{N}) \otimes \mathbf{C}) \subset (\delta \otimes \iota)(\mathfrak{N} \,\bar{\otimes}\, \mathcal{R}(G)) \ .$$

Since $V_G \in \mathcal{R}(G) \,\bar{\otimes}\, L^\infty(G)$ and $W_G \in L^\infty(G) \,\bar{\otimes}\, \mathcal{R}(G)$, we have

$$(3.1) \qquad \mathrm{Ad}_{(1\otimes V_G^*)}(\alpha(\mathfrak{M})' \otimes \mathbf{C}) \subset \alpha(\mathfrak{M})' \,\bar{\otimes}\, L^\infty(G)$$

$$(3.2) \qquad \mathrm{Ad}_{(1\otimes W_G)}(\delta(\mathfrak{N})' \otimes \mathbf{C}) \subset \delta(\mathfrak{N})' \,\bar{\otimes}\, \mathcal{R}(G) \ .$$

Besides, V_G^* and W_G satisfy the associativity conditions (I.4.11) and (I.4.6).

Definition 3.1. We define an action β of G on $\alpha(\mathfrak{M})'$ with respect to $\mathcal{R}(G)'$ and a co-action ε of G on $\delta(\mathfrak{N})'$ by

$$(3.3) \qquad \beta(x) = \mathrm{Ad}_{1\otimes V_G^*}(x \otimes 1) \ , \qquad x \in \alpha(\mathfrak{M})'$$

$$(3.4) \qquad \varepsilon(y) = \mathrm{Ad}_{1\otimes W_G}(y \otimes 1) \ , \qquad y \in \delta(\mathfrak{N})' \ .$$

For a closed subgroup H of G we denote

$$\mathcal{L}^\infty(H\backslash G) = L^\infty(G) \cap \lambda(H)'$$

$$\mathcal{L}^\infty(G/H) = L^\infty(G) \cap \rho(H)'$$

$$\mathfrak{M} \times_\alpha H = \alpha(\mathfrak{M}) \vee (\mathbf{C} \otimes \rho(H)'')$$

$$\mathfrak{N} \times_\delta (H\backslash G) = \delta(\mathfrak{N}) \vee (\mathbf{C} \otimes \mathcal{L}^\infty(H\backslash G)) \ .$$

Now, we shall show an imprimitivity theorem:

Theorem 3.2. (a) (Assume that H is normal). The restriction of β to $(\mathfrak{M} \times_\alpha H)'$ is an action of G with respect to $\mathcal{R}(G)'$ and

$$(3.5) \qquad (\mathfrak{M} \times_\alpha G)' = ((\mathfrak{M} \times_\alpha H)')^\beta \ ;$$

$$(3.6) \qquad (\mathfrak{M} \times_\alpha H)' = (\mathfrak{M} \times_\alpha G)' \vee (\mathbf{C} \otimes \mathcal{L}^\infty(G/H)) \ .$$

(b) The restriction of ε to $(\mathfrak{n} \times_\delta (H\backslash G))'$ is a co-action of G and

(3.7) $$(\mathfrak{n} \times_\delta G)' = ((\mathfrak{n} \times_\delta (H\backslash G))')^\varepsilon$$

(3.8) $$(\mathfrak{n} \times_\delta (H\backslash G))' = (\mathfrak{n} \times_\delta G)' \vee (\mathbf{C} \otimes \lambda(H)'') \quad .$$

Proof. (a) Put $\mathfrak{m}_0 = (\mathfrak{m} \times_\alpha H)'$. Since H is normal and $V_G \in \mathcal{R}(G) \bar{\otimes} L^\infty(G)$, we have

$$\mathrm{Ad}_{1\otimes V_G}(1 \otimes \rho(r) \otimes f) \in \mathbf{C} \otimes \rho(H)'' \bar{\otimes} L^\infty(G) \;, \qquad r \in H \;.$$

Combining this with (3.1), we have $\beta(\mathfrak{m}_0) \subset \mathfrak{m}_0 \bar{\otimes} L^\infty(G)$. Therefore the restriction of β to $\mathfrak{m}_0$ is an action of G with respect to $\mathcal{R}(G)'$.

Next we shall show (3.5). Since $V_G \in \mathcal{R}(G) \bar{\otimes} L^\infty(G)$, $(\mathfrak{m} \times_\alpha G)' \subset ((\mathfrak{m} \times_\delta H)')^\beta$. It suffices to show the reverse inclusion. If $x \in (\alpha(\mathfrak{m})')^\beta$, then $x \otimes 1 = \mathrm{Ad}_{1\otimes V_G^*}(x \otimes 1)$ belongs to the commutant of $\mathbf{C} \otimes \mathcal{R}(G) \bar{\otimes} \mathcal{R}(G)$, for

(3.9) $$(\mathbf{C} \otimes \mathcal{R}(G)) \vee V_G^*(\mathbf{C} \otimes \mathcal{R}(G))V_G = \mathcal{R}(G) \bar{\otimes} \mathcal{R}(G) \;.$$

Therefore $x \in \alpha(\mathfrak{m})' \cap (\mathbf{C} \otimes \mathcal{R}(G))' = (\mathfrak{m} \times_\alpha G)'$. Thus

$$((\mathfrak{m} \times_\alpha H)')^\beta \subset (\alpha(\mathfrak{m})')^\beta \subset (\mathfrak{m} \times_\alpha G)' \quad .$$

Finally we shall show (3.6). As $\mathfrak{m}_0$ commutes with $\mathbf{C} \otimes \rho(H)''$ and as $V_G^*(1 \otimes \rho(r))V_G = \rho(r) \otimes \rho(r)$, $\beta(\mathfrak{m}_0)$ is contained in $\mathfrak{m}_0 \bar{\otimes} \mathcal{L}^\infty(G/H)$. Let Φ be an isomorphism of $L^\infty(G/H)$ onto $\mathcal{L}^\infty(G/H)$ with $\Phi(f) = f \circ \varphi$ for $f \in L^\infty(G/H)$, where φ is the canonical map of G on G/H. Here we set

$$\dot{\beta}(x) = (\iota \otimes \iota \otimes \Phi^{-1}) \circ \beta(x) \;, \qquad x \in \mathfrak{m}_0 \;.$$

Then $\dot{\beta}$ is an action of G/H on $\mathfrak{m}_0$ such that $\dot{\beta}_{\dot{t}} = \beta_t$ with $\dot{t} = tH$. We set

$$\pi(f) = 1 \otimes \Phi(f) \;, \qquad f \in L^\infty(G/H) \;.$$

Since $\mathbf{C} \otimes \mathcal{L}^\infty(G/H) \subset \mathfrak{m}_0$, π is an isomorphism of $L^\infty(G/H)$ into $\mathfrak{m}_0$ satisfying

(3.10) $$\dot{\beta}_{\dot{t}} \circ \pi(f) = \beta_t(1 \otimes (f \circ \varphi)) = \pi({}_{\dot{t}^{-1}}f) \;,$$

where
$$ {}_{\dot{t}^{-1}}f(\dot{s}) = f(\dot{s}\dot{t}^{-1}) \quad .$$

Therefore, by Theorem 2.2, $\mathfrak{m}_0$ is generated by $(\mathfrak{m}_0)^{\dot{\beta}}$ and $\pi(L^\infty(G/H))$. Since $(\mathfrak{m}_0)^{\dot{\beta}} = (\mathfrak{m}_0)^\beta$ and $\pi(L^\infty(G/H)) = \mathbf{C} \otimes \mathcal{L}^\infty(G/H)$, $\mathfrak{m}_0$ is generated by $(\mathfrak{m} \times_\alpha G)'$ and $\mathbf{C} \otimes \mathcal{L}^\infty(G/H)$.

(b) Put $\mathfrak{n}_o = (\mathfrak{n} \times_\delta (H\backslash G))'$. Since $W_G \in L^\infty(G) \bar{\otimes} \mathcal{R}(G)$,

$$\mathrm{Ad}_{1\otimes W_G}(\mathbb{C} \otimes \mathcal{L}^\infty(H\backslash G) \otimes \mathbb{C}) = \mathbb{C} \otimes \mathcal{L}^\infty(H\backslash G) \otimes \mathbb{C} .$$

Combining this with (3.2), we have $\varepsilon(\mathfrak{n}_o) \subset \mathfrak{n}_o \bar{\otimes} \mathcal{R}(G)$. Therefore the restriction of ε to $\mathfrak{n}_o$ is a co-action of G.

Next we shall show (3.7). Since $W_G \in L^\infty(G) \bar{\otimes} \mathcal{R}(G)$, $(\mathfrak{n}\times_\delta G)' \subset ((\mathfrak{n}\times_\delta (H\backslash G))')^\varepsilon$, it suffices to show the reverse inclusion. If $y \in (\delta(\mathfrak{n})')^\varepsilon$, then $y \otimes 1 = \mathrm{Ad}_{1\otimes W_G}(y \otimes 1)$ belongs to the commutant of $\mathbb{C} \otimes L^\infty(G) \bar{\otimes} L^\infty(G)$, for

$$(3.11) \qquad (\mathbb{C} \otimes L^\infty(G)) \vee W_G(\mathbb{C} \otimes L^\infty(G))W_G^* = L^\infty(G) \bar{\otimes} L^\infty(G) \quad .$$

Therefore, $y \in \delta(\mathfrak{n})' \cap (\mathbb{C} \otimes L^\infty(G))' = (\mathfrak{n} \times_\delta G)'$. Thus,

$$((\mathfrak{n} \times_\delta (H\backslash G))')^\varepsilon \subset (\delta(\mathfrak{n})')^\varepsilon \subset (\mathfrak{n} \times_\delta G)' .$$

Finally we shall show (3.8). As $\mathfrak{n}_o$ commutes with $\mathbb{C} \otimes \mathcal{L}^\infty(H\backslash G)$, $\varepsilon(\mathfrak{n}_o)$ is contained in $\mathfrak{n}_o \bar{\otimes} \rho(H)''$. Let Ψ be an isomorphism of $\rho^H(H)''$ onto $\rho(H)''$ with $\Psi(\rho^H(r)) = \rho(r)$, where ρ^H is the right regular representation of H on $L^2(H)$. Here we set

$$\varepsilon^H(y) = (\iota \otimes \iota \otimes \Psi^{-1}) \circ \varepsilon(y) , \qquad y \in \mathfrak{n}_o .$$

Then ε^H is a co-action of H on $\mathfrak{n}_o$. Since $\mathbb{C} \otimes \lambda(H) \subset \mathfrak{n}_o$ and

$$(3.12) \qquad \begin{aligned} \varepsilon^H(1 \otimes \lambda(r)) &= (\iota \otimes \iota \otimes \Psi^{-1}) \circ \varepsilon(1 \otimes \lambda(r)) \\ &= (\iota \otimes \iota \otimes \Psi^{-1})(1 \otimes \lambda(r) \otimes \rho(r)) = 1 \otimes \lambda(r) \otimes \rho^H(r) , \end{aligned}$$

it follows from Theorem 2.1 that $\mathfrak{n}_o$ is generated by $(\mathfrak{n}_o)^{\varepsilon^H}$ and $\mathbb{C} \otimes \lambda(H)$. Since $(\mathfrak{n}_o)^{\varepsilon^H} = (\mathfrak{n}_o)^\varepsilon$, $\mathfrak{n}_o$ is generated by $(\mathfrak{n} \times_\delta G)'$ and $\mathbb{C} \otimes \lambda(H)$. Q.E.D.

<u>Corollary</u> 3.3. (a) The action β is dual on $\alpha(\mathfrak{m})'$ and

$$(3.13) \qquad (\alpha(\mathfrak{m})')^\beta = (\mathfrak{m} \times_\alpha G)'$$

(b) The co-action ε is dual on $\delta(\mathfrak{n})'$ and

$$(3.14) \qquad (\delta(\mathfrak{n})')^\varepsilon = (\mathfrak{n} \times_\delta G)'$$

<u>Proof</u>. (a) By (3.10), β is dual on $\alpha(\mathfrak{m})'$.
(b) By (3.12), ε is dual on $\delta(\mathfrak{n})'$.

Corollary 3.4. (a) If α is implemented by a unitary u in $\mathcal{L}(\mathfrak{H})\overline{\otimes}L^\infty(G)$ satisfying the associativity condition

$$(u\otimes 1)(\iota\otimes\sigma)(u\otimes 1) = (\iota\otimes\alpha_G)(u)$$

such that $\alpha(x) = u(x\otimes 1)u^*$, then

$$(3.15)\qquad (\mathfrak{M}\times_\alpha G)' = (\mathfrak{M}'\otimes\mathbb{C})\vee u(\mathbb{C}\otimes\mathcal{R}(G)')u^* .$$

(b) If δ is implemented by a unitary w in $\mathcal{L}(\mathfrak{K})\overline{\otimes}\mathcal{R}(G)$ satisfying the associativity condition

$$(w^*\otimes 1)(\iota\otimes\sigma)(w^*\otimes 1) = (\iota\otimes\delta_G)(w^*)$$

such that $\delta(y) = w^*(y\otimes 1)w$, then

$$(3.16)\qquad (\mathfrak{N}\times_\delta G)' = (\mathfrak{N}'\otimes\mathbb{C})\vee w^*(\mathbb{C}\otimes L^\infty(G))w .$$

Proof. (a) By the associativity condition, $u(1\otimes\lambda(r))u^* = u(r)\otimes\lambda(r)$. According to Theorem 1.2, $\mathfrak{M}\times_\alpha G$ is $(\mathfrak{M}\overline{\otimes}\mathcal{L}(L^2(G)))^{\tilde\alpha}$. Since $\tilde\alpha_t = \alpha_t\otimes\lambda_t$, $\mathfrak{M}\times_\alpha G$ is the intersection of $\mathfrak{M}\overline{\otimes}\mathcal{L}(L^2(G))$ with the commutant of $u(t)\otimes\lambda(t)$, $t\in G$. Therefore $(\mathfrak{M}\times_\alpha G)'$ is generated by $\mathfrak{M}'\otimes\mathbb{C}$ and $u(\mathbb{C}\otimes\mathcal{R}(G)')u^*$.

(b) By virtue of Theorem 1.2, we have only to show that, for each $y\in\mathfrak{N}\overline{\otimes}\mathcal{L}(L^2(G))$,

$$\tilde\delta(y) = y\otimes 1 \qquad \text{if and only if} \qquad y\in\pi(L^\infty(G))' ,$$

where $\pi(f) = w^*(1\otimes f)w$ for $f\in L^\infty(G)$. The associativity condition implies that

$$(3.17)\qquad (1\otimes W_G)(\iota\otimes\sigma)(w^*\otimes 1) = (w^*\otimes 1)(1\otimes W_G)(w\otimes 1) ,$$

so that

$$\tilde\delta(y) = \mathrm{Ad}_{(w^*\otimes 1)(1\otimes W_G)(w\otimes 1)}(y\otimes 1) .$$

Therefore $y\in\pi(L^\infty(G))'$ implies $\tilde\delta(y) = y\otimes 1$.

Conversely, the associativity condition implies that

$$(1\otimes W_G)(w^*\otimes 1)(\iota\otimes\sigma)(w^*\otimes 1) = (w^*\otimes 1)(1\otimes W_G)$$

and hence that

$$\mathrm{Ad}_{1\otimes W_G}(\pi(f)\otimes 1) = \pi(f)\otimes 1 .$$

Therefore, by (3.14), we have $\pi(f) \in (\mathcal{M} \times_\delta G)'$ and hence $\pi(f)$ commutes with y for each $f \in L^\infty(G)$ by Theorem 1.2. Q.E.D.

The same type of formula as (3.17) is obtained from the associativity condition on u:

$$(1 \otimes V'_G)(\iota \otimes \sigma)(u \otimes 1) = (u \otimes 1)(1 \otimes V'_G)(u^* \otimes 1) \quad . \tag{3.18}$$

NOTES

The action β, Definition 3.1, is introduced by Landstad, [43]. The imprimitivity theorem, Theorem 3.2 and Corollary 3.3.a, are due to Nakagami [47]. Corollary 3.4 is due to [16,33,43,47].

CHAPTER III.

INTEGRABILITY AND DOMINANCE

Introduction. As in the case of actions, [14], integrable co-actions play a crucial role in the analysis of the crossed product. To do this, we shall prepare in §1 elementary properties of the operator valued weight associated with the Plancherel weight ψ_G on $\mathcal{R}(G)$ as well as the Haar measure μ_G. For an action α of G on $\mathfrak{M}$, the $\mathfrak{M}^\alpha$-valued weight $\mathcal{E}_\alpha$ is defined as the integral with respect to μ_G and for a co-action δ of G on $\mathfrak{N}$, the $\mathfrak{N}^\delta$-valued weight $\mathcal{E}_\delta$ is then the integral with respect to ψ_G. The integrability of α (resp. δ) is defined by the semi-finiteness of $\mathcal{E}_\alpha$ (resp. $\mathcal{E}_\delta$) in §2. Making use of $\mathcal{E}_\delta$, we shall show there that an integrable co-action δ on a standard von Neumann algebra $\{\mathfrak{N},\mathfrak{K}\}$ is implemented by a representation of $C_\infty(G)$ on $\mathfrak{K}$ which is identified with a certain unitary in $\mathcal{L}(\mathfrak{K}) \bar{\otimes} \mathcal{R}(G)$ - this identification is discussed in Appendix. It is, however, conjectured that the implementability for a standard von Neumann algebra holds without the integrability assumption for δ.

In §3, integrable actions (resp. co-actions) are characterized as reduced actions (resp. co-actions) of the second dual actions (resp. co-action), which is also equivalent to the point spectrum property of all reduced actions (resp. co-actions), (3.1) and (3.4).

In §4, it will be shown that among integrable actions (resp. co-actions) there is a unique, up to equivalence, largest one, which is dual and of infinite multiplicity. Such an action (resp. a co-action) is called dominant.

§1. Operator valued weights

Let μ_G (resp. μ'_G) be the faithful, semi-finite, normal weight on $L^\infty(G)$ given by integration with respect to a right (resp. left) invariant Haar measure. We normalize μ_G and μ'_G so that $\mu'_G(f) = \mu_G(\Delta f)$, $f \in \mathcal{K}(G)$.

Given an action α of G on $\mathfrak{M}$, we set, for each $f \in L^1(G)$,

$$\text{(1.1)} \qquad \begin{aligned} \langle \alpha_f(x), \omega \rangle &= \langle \alpha(x), \omega \otimes f \rangle = \int_G \langle \alpha_t(x), \omega \rangle f(t) dt \\ &= \left\langle \int_G f(t)\alpha_t(x) dt, \omega \right\rangle , \quad \omega \in \mathfrak{M}_* . \end{aligned}$$

Let $F = \{g \in \mathcal{K}(G) : 0 \leq g \leq \Delta\}$. If $x \in \mathfrak{M}_+$, then $\{\alpha_g(x) : g \in F\}$ is an increasing net in $\mathfrak{M}_+$. We define an $\mathfrak{M}$-valued weight $\mathcal{E}_\alpha$ by

$$\text{(1.2)} \qquad \mathcal{E}_\alpha(x) = \sup\{\alpha_g(x) : g \in F\} , \quad x \in \mathfrak{M}_+ ,$$

if the right hand side exists in $\mathfrak{M}$. Since we have

$$\alpha_s \circ \alpha_f = \Delta(s)\alpha_{\lambda_s f} , \quad f \in \mathcal{K}(G) ;$$

$$\Delta(s)\lambda_s F = F ,$$

$\mathcal{E}_\alpha(x)$ falls in $\mathfrak{M}^\alpha$ if it exists. As in the case of numerical valued weights, we consider

$$\mathfrak{q}_\alpha = \{x \in \mathfrak{M} : \alpha_g(x^*x) , \ g \in F, \ \text{is bounded}\} .$$

It then follows that $\mathfrak{q}_\alpha$ is a left ideal of $\mathfrak{M}$. Put

$$\mathfrak{p}_\alpha = \mathfrak{q}_\alpha^* \mathfrak{q}_\alpha .$$

It is easily seen that $\mathcal{E}_\alpha$ is extended to a linear map of $\mathfrak{p}_\alpha$ into $\mathfrak{M}^\alpha$, which will be denoted by $\dot{\mathcal{E}}_\alpha$. As a special case, we have

$$\text{(1.3)} \qquad \dot{\mathcal{E}}_{\alpha_G}(f) = \mu'_G(f)1 , \quad f \in L^\infty(G) \cap L^1(G, d's) .$$

Proposition 1.1.

$$\text{(1.4)} \qquad \mathcal{E}_\alpha(x_i) \nearrow \mathcal{E}_\alpha(x) \ \text{ if } \ x_i \nearrow x \ \text{ in } \ \mathfrak{M}_+ ;$$

(1.5) $$\alpha(\dot{\mathcal{E}}_\alpha(x)) = \dot{\mathcal{E}}_\alpha(x) \otimes 1 , \quad x \in \mathfrak{p}_\alpha ;$$

(1.6) $$\mathfrak{m}^\alpha \mathfrak{p}_\alpha \mathfrak{m}^\alpha \subset \mathfrak{p}_\alpha \quad \text{and} \quad \dot{\mathcal{E}}_\alpha(axb) = a\dot{\mathcal{E}}_\alpha(x)b , \quad a,b \in \mathfrak{m}^\alpha , \quad x \in \mathfrak{p}_\alpha ;$$

(1.7) the map : $x \in \mathfrak{M} \to \dot{\mathcal{E}}_\alpha(a^*xb) \in \mathfrak{M}^\alpha$ is σ-weakly continuous for each $a,b \in \mathfrak{q}_\alpha$.

The proof is very similar to that for the corresponding properties of weights, so we leave it to the reader.

We now want to dualize the above construction of $\mathcal{E}_\alpha$. For this, we must first dualize the notion of Haar measure. So we shall define the Plancherel weight on $\mathfrak{R}(G)$. We need the Tomita algebra which gives rise to $\mathfrak{R}(G)$. We introduce a Tomita algebra structure in $\mathcal{K}(G)$ as follows:

(1.8) $$\begin{cases} f*g(t) = \int_G f(ts^{-1})g(s)ds , \quad f^\#(t) = \overline{f(t^{-1})} ; \\ (\Delta^{i\alpha}f)(t) = \Delta(t)^{i\alpha}f(t) , \quad \alpha \in C ; \ (f|g) = \int f(t)\overline{g(t)}\, dt . \end{cases}$$

We then have

(1.9) $$\begin{cases} f^\flat(t) = \Delta(t)\overline{f(t^{-1})} , \quad f \in \mathcal{K}(G) ; \\ \pi_\ell(f) = \lambda(\Delta^{\frac{1}{2}}f) ; \\ \pi_r(f) = \rho(\Delta f^\vee) , \end{cases}$$

where λ and ρ are of course the left and right regular representations of G and

(1.10) $$\lambda(f) = \int_G f(s)\lambda(s)ds , \quad \rho(f) = \int_G f(s)\rho(s)ds .$$

It then follows that our von Neumann algebra $\mathfrak{R}(G)$ is the right von Neumann algebra $\mathfrak{R}_r(\mathcal{K}(G))$ of the Tomita algebra $\mathcal{K}(G)$ and its commutant $\mathfrak{R}'(G)$ is the left von Neumann algebra $\mathfrak{R}_\ell(\mathcal{K}(G))$. Thus, we set

$$\mathfrak{A} = \mathcal{K}(G)' \quad \text{and} \quad \mathfrak{A}' = \mathcal{K}(G)''$$

and get a full right Hilbert algebra $\mathfrak{A}$ with $\mathfrak{R}_r(\mathfrak{A}) = \mathfrak{R}(G)$ and a full left Hilbert algebra $\mathfrak{A}'$ with $\mathfrak{R}_\ell(\mathfrak{A}') = \mathfrak{R}'(G)$. The modular unitary involution J associated with $\mathcal{K}(G)$(or $\mathfrak{A}$) is given by:

(1.11) $$J\xi(s) = \Delta(s)^{\frac{1}{2}}\, \overline{\xi(s^{-1})} , \quad \xi \in L^2(G) .$$

The canonical weight ψ_G on $\mathcal{R}_r(\mathfrak{A}) = \mathcal{R}(G)$ is then given by:

(1.12) $$\psi_G(\pi_r(g)^*\pi_r(f)) = (f|g) = (f*g^b)(e) = (g^\# * f)(e) ,$$

where e is the unit of G. We shall call ψ_G the <u>Plancherel</u> <u>weight</u> on $\mathcal{R}(G)$. The corresponding modular automorphism group is then given by

(1.13) $$\sigma_t(x) = \Delta^{-it} x \Delta^{it} , \quad x \in \mathcal{R}(G) .$$

Set

(1.14) $$\Psi = \{\varphi \in \mathcal{R}(G)_*^+ : (1+\varepsilon)\varphi \le \psi_G \text{ for some } \varepsilon = \varepsilon_\varphi > 0\} .$$

We then have

(1.15) $$\psi_G(x) = \sup\{\varphi(x) : \varphi \in \Psi\} .$$

Suppose now we have a co-action δ of G on $\mathfrak{n}$. By (I.2.18), we define, for each $\varphi \in \mathcal{R}(G)_* = A(G)$, a linear transformation δ_φ on $\mathfrak{n}$. It then follows that $\{\delta_\varphi : \varphi \in \Psi\}$ is upward directed. We set

(1.16) $$\mathcal{E}_\delta(x) = \sup\{\delta_\varphi(x) : \varphi \in \Psi\} , \quad x \in \mathfrak{n}_+ .$$

Of course, $\mathcal{E}_\delta(x)$ does not necessarily exist. So we mean that $\mathcal{E}_\delta(x)$ is defined only for those $x \in \mathfrak{n}_+$ such that the right hand side of (1.16) exists. We then have, for each $\omega \in \mathfrak{n}_*^+$,

(1.17) $$\begin{aligned}\langle \mathcal{E}_\delta(x), \omega\rangle &= \sup\{\langle \delta(x), \omega \otimes \varphi\rangle : \varphi \in \Psi\} \\ &= \langle \delta(x), \omega \otimes \psi_G\rangle , \qquad x \in \mathfrak{n}_+ .\end{aligned}$$

As before we set

(1.18) $$\mathfrak{q}_\delta = \{x \in \mathfrak{n} : \mathcal{E}_\delta(x^*x) \text{ exists}\} \text{ and } \mathfrak{p}_\delta = \mathfrak{q}_\delta^* \mathfrak{q}_\delta .$$

We then extend $\mathcal{E}_\delta$ to a linear map $\dot{\mathcal{E}}_\delta$ of $\mathfrak{p}_\delta$ into $\mathfrak{n}$. As a special case, we have the following formula which requires a little bit of proving:

<u>Lemma</u> 1.2. For the co-action $\delta = \delta_G$ of G on $\mathcal{R}(G)$,

(1.19) $$\dot{\mathcal{E}}_\delta(x) = \psi_G(x)1 , \quad x \in \mathfrak{p}_\delta = \mathfrak{m}_{\psi_G} .$$

<u>Proof</u>. We identify first $\mathcal{R}(G)_*$ with the algebra $A(G)$ of functions on G which are of the form: $g^\# * f$, $f,g \in L^2(G)$, under the correspondence: $\omega_{f,g} \in \mathcal{R}(G)_* \longleftrightarrow g^\# * f \in A(G)$. It follows from the construction of ψ_G that if $\eta_i \in \mathcal{K}(G)$ is a net such that $\pi_\ell(\eta_i)^*\pi_\ell(\eta_i) \nearrow 1$ then $\omega_{\eta_i}(x) = (x\eta_i|\eta_i) \nearrow \psi_G(x)$

for every $x \in \mathcal{R}(G)_+$. Suppose $\varphi \in A(G)_+$ is a state. For any $f \in A(G)_+$ we put $g = f\varphi$. Then

$$\langle \delta(\rho(f)), \varphi \otimes \omega_{\eta_i} \rangle = \langle \rho(g), \omega_{\eta_i} \rangle .$$

Since $f, g \in A(G)_+$, we have $\rho(f) \geq 0$ and $\rho(g) \geq 0$ and hence

$$(1.20) \qquad (\varphi \otimes \psi_G) \circ \delta(\rho(f)) = g(e) = f(e) = \psi_G(\rho(f)) .$$

Now let $\{\sigma_t : t \in \mathbb{R}\}$ be the modular automorphism group of $\mathcal{R}(G)$ associated with ψ_G. We then have

$$(1.21) \qquad \sigma_t(\rho(s)) = \Delta(s)^{it} \rho(s) , \quad s \in G, \quad t \in \mathbb{R} ;$$

hence

$$(1.22) \qquad \delta \circ \sigma_t = (\sigma_t \otimes \iota) \circ \delta = (\iota \otimes \sigma_t) \circ \delta .$$

Thus, if we set $\psi = (\psi_G \otimes \varphi) \circ \delta$ for a state $\varphi \in A(G)$, then $\psi \circ \sigma_t = \psi$, so that ψ and ψ_G commute in the sense of [80]. We claim now that $\psi = \psi_G$ if $\varphi \in A(G)$ is a state. To this end, we need only to check the equality on a σ-weakly dense part, see [80,81]. Since the set of all $\xi * \eta^\flat$ with $\xi, \eta \in \mathfrak{A}$ is dense in $L^1(G)$, the set of all $\rho(\xi * \eta^\flat)$ with $\xi, \eta \in \mathfrak{A}$ is σ-weakly dense in $\mathcal{R}(G)$. Therefore, we finally get $\psi_G = (\psi_G \otimes \varphi) \circ \delta$ for any normal state φ on $\mathcal{R}(G)$. By linearity, we have

$$(1.23) \qquad \varphi(1)\psi_G = (\psi_G \otimes \varphi) \circ \delta ,$$

which means precisely (1.19). Q.E.D.

<u>Corollary</u> 1.3. In the general case, we have

$$(1.24) \qquad \dot{\mathcal{E}}_\delta(\mathfrak{p}_\delta) \subset \mathfrak{h}^\delta .$$

<u>Proof</u>. For each $x \in \mathfrak{p}_\delta^+$, $\varphi \in \mathfrak{h}_*^+$ and $\psi \in \mathcal{R}(G)_*^+$, we have

$$\langle \delta \circ \mathcal{E}_\delta(x), \varphi \otimes \psi \rangle = \langle \mathcal{E}_\delta(x), \delta_*(\varphi \otimes \psi) \rangle$$

$$= \langle \delta(x), \delta_*(\varphi \otimes \psi) \otimes \psi_G \rangle = \langle (\delta \otimes \iota) \circ \delta(x), \varphi \otimes \psi \otimes \psi_G \rangle$$

$$= \langle (\iota \otimes \delta_G) \circ \delta(x), \varphi \otimes \psi \otimes \psi_G \rangle$$

$$= \langle \delta(x), \varphi \otimes (\psi \otimes \psi_G) \circ \delta_G \rangle = \langle 1, \psi \rangle \langle \delta(x), \varphi \otimes \psi_G \rangle, \ (1.23) \ ,$$

$$= \langle 1, \psi \rangle \langle \mathcal{E}_\delta(x), \varphi \rangle = \langle \mathcal{E}_\delta(x) \otimes 1, \varphi \otimes \psi \rangle \ ,$$

which means by definition that $\mathcal{E}_\delta(x) \in \mathfrak{n}^\delta$. Q.E.D.

Proposition 1.4. With a co-action δ of G on $\mathfrak{n}$, we have the following:

(1.25) $\mathcal{E}_\delta(\sup x_i) = \sup \mathcal{E}_\delta(x_i)$ for any bounded increasing net $\{x_i\}$ in $\mathfrak{n}_+$;

(1.26) $$\mathfrak{n}^\delta \mathfrak{p}_\delta \mathfrak{n}^\delta \subset \mathfrak{p}_\delta \ ;$$

(1.27) $$\dot{\mathcal{E}}_\delta(axb) = a\dot{\mathcal{E}}_\delta(x)b \ , \quad a,b \in \mathfrak{n}^\delta, \quad x \in \mathfrak{p}_\delta \ ;$$

(1.28) the map : $x \in \mathfrak{n} \to \dot{\mathcal{E}}_\delta(a^* x a)$ is σ-weakly continuous for every $a \in \mathfrak{q}_\delta$.

Proof . The normality, (1.25), of $\mathcal{E}_\delta$ follows directly from its construction. (1.26) and (1.27): If $a \in \mathfrak{n}^\delta$ and $x \in \mathfrak{n}$, then we have for each $\varphi \in \mathfrak{n}_*^+$,

$$\langle \delta(a^* x^* x a), \varphi \otimes \psi_G \rangle = \langle (a^* \otimes 1)\delta(x^* x)(a \otimes 1), \varphi \otimes \psi_G \rangle$$

$$= \langle \delta(x^* x), (a\varphi a^*) \otimes \psi_G \rangle \ ;$$

hence if $x \in \mathfrak{q}_\delta$, then we have $xa \in \mathfrak{q}_\delta$ and

$$\langle a^* \mathcal{E}_\delta(x^* x) a, \varphi \rangle = \langle \delta(x^* x), a\varphi a^* \otimes \psi_G \rangle = \langle \mathcal{E}_\delta(a^* x^* x a), \varphi \rangle \ ;$$

thus (1.26) and (1.27) both follow.

Claim (1.28) is routine. Q.E.D.

We state here a formula which has been implicitly used, while the proof itself is routine:

(1.29)
$$\delta_{\varphi\psi} = \delta_{\varphi}\delta_{\psi}\ , \quad \varphi,\psi \in A(G)\ .$$

Lemma 1.5. If $\{\psi_i\}$ is a net in $A(G) \cap \mathcal{K}(G)_+$ such that $\{\Delta\psi_i\}$ is a bounded approximate identity in the Banach algebra $L^1(G)$, then

(1.30)
$$\delta(x) = \lim_i \int_G \delta_{\rho(s)^*\psi_i}(x) \otimes \rho(s)ds$$

in the σ-weak topology for every x of the form:

$$x = \delta_{\varphi}(y)\ , \quad y \in \mathfrak{N}\ , \quad \varphi \in A(G) \cap \mathcal{K}(G)\ .$$

Proof. If $\varphi,\psi \in A(G) \cap \mathcal{K}(G)$, then the $A(G)$-valued function: $s \in G \to \varphi(\rho(s)^*\psi) \in A(G)$ has a compact support $K = (\operatorname{supp}\psi)^{-1}(\operatorname{supp}\varphi)$, so that the function:

$$s \in G \to \delta_{\varphi(\rho(s)^*\psi)}(y) \otimes \rho(s) \in \mathfrak{N}\,\overline{\otimes}\,\mathcal{R}(G)$$

is continuous and supported by K. Hence the following integral makes sense:

$$\int_G \delta_{\varphi(\rho(s)^*\psi_i)}(y) \otimes \rho(s)ds = z_i\ .$$

For each $f \in A(G)$, we compute

$$\begin{aligned}
\langle\rho(t), \int_G f(s)\varphi(\rho(s)^*\psi)ds\rangle &= \int_G \langle\rho(t),\varphi(\rho(s)^*\psi)\rangle f(s)ds \\
&= \int_G \varphi(t)\psi(ts^{-1})f(s)ds \quad = \varphi(t)\int_G \psi(s^{-1})f(st)ds \\
&= \varphi(t)\int_G \Delta(s)\psi(s)f(s^{-1}t)ds \quad = \varphi(t)\int_G \Delta(s)\psi(s)(\lambda_s f)(t)ds \\
&= \varphi(t)\lambda_{\Delta\psi}(f)(t) \quad = \langle\rho(t),\varphi(\lambda_{\Delta\psi}(f))\rangle\ ;
\end{aligned}$$

with $\omega \in \mathfrak{N}_*$,

$$\langle z_i, \omega \otimes f\rangle = \int_G \left\langle \delta_{\varphi(\rho(s)^*\psi_i)}(y), \omega\right\rangle f(s)ds$$

$$= \int_G \langle \delta(y), \omega \otimes \varphi(\rho(s)^*\psi_i)\rangle f(s)ds$$

$$= \left\langle \delta(y), \omega \otimes \int_G f(s)\varphi(\rho(s)^*\psi_i)ds\right\rangle$$

$$= \left\langle \delta(y), \omega \otimes \varphi\lambda_{\Delta\psi_i}(f)\right\rangle = \left\langle \delta \circ \delta_\varphi(y)', \left(\iota \otimes \lambda_{\Delta\psi_i}\right)(\omega \otimes f)\right\rangle .$$

Hence we get, for every $\Phi \in (\mathfrak{n} \otimes \mathcal{R}(G))_*$,

$$\langle z_i, \Phi\rangle = \left\langle \delta \circ \delta_\varphi(y), \left(1 \otimes \lambda_{\Delta\psi_i}\right)\Phi\right\rangle = \left\langle \delta(x), \left(1 \otimes \lambda_{\Delta\psi_i}\right)\Phi\right\rangle$$

so that

$$\lim_i \langle z_i, \Phi\rangle = \lim_i \left\langle \delta(x), \left(1 \otimes \lambda_{\Delta\psi_i}\right)\Phi\right\rangle = \langle \delta(x), \Phi\rangle . \qquad \text{Q.E.D.}$$

<u>Proposition</u> 1.6. a) If α is an action of G on $\mathfrak{m}$, then we have, for the dual co-action $\hat{\alpha}$ of G on $\mathfrak{m} \times_\alpha G$, the following:

(1.31) $\qquad C \otimes \mathfrak{m}_{\psi_G} \subset \mathfrak{p}_{\hat{\alpha}}$ and $\dot{\mathcal{E}}_{\hat{\alpha}}(1 \otimes x) = \psi_G(x)1, \quad x \in \mathfrak{m}_{\psi_G}$;

(1.32) $$\dot{\mathcal{E}}_{\hat{\alpha}}(\mathfrak{p}_{\hat{\alpha}}) = (\mathfrak{m} \times_\alpha G)^{\hat{\alpha}} .$$

b) If δ is a co-action of G on $\mathfrak{n}$, then we have the following for the dual action $\hat{\delta}$ of G on $\mathfrak{n} \times_\delta G$:

(1.33) $\quad C \otimes (L^\infty(G) \cap L^1(G)) \subset \mathfrak{p}_{\hat{\delta}}$ and $\dot{\mathcal{E}}_{\hat{\delta}}(1 \otimes g) = \mu_G(g)1, \quad g \in L^1(G) \cap L^\infty(G)$;

(1.34) $$\dot{\mathcal{E}}_{\hat{\delta}}(\mathfrak{p}_{\hat{\delta}}) = (\mathfrak{n} \times_\delta G)^{\hat{\delta}} ;$$

(1.35) $\qquad$ if $g \in \mathcal{K}(G)$, then with $\varphi = g^\# * g \in A(G)$,

$$\dot{\mathcal{E}}_{\hat{\delta}}((1 \otimes g)^*\delta(y)(1 \otimes g)) = \delta \circ \delta_\varphi(y), \quad y \in \mathfrak{n} .$$

<u>Proof</u> (1.31): By definition, the restriction of $\hat{\alpha}$ to $\mathbf{C} \otimes \mathcal{R}(G)$ coincides with the canonical co-action δ_G, i.e., $\hat{\alpha}(1 \otimes x) = 1 \otimes \delta_G(x)$, $x \in \mathcal{R}(G)$. Hence our assertion follows from (1.19).

(1.32): By (1.24), $\dot{\mathcal{E}}_{\hat{\alpha}}(\mathfrak{p}_{\hat{\alpha}}) \subset (\mathfrak{M} \times_\alpha G)^{\hat{\alpha}}$. If $a \in (\mathfrak{M} \times_\alpha G)^{\hat{\alpha}}$ and $x \in \mathfrak{m}_{\psi_G}$, then $a(1 \otimes x) \in \mathfrak{p}_{\hat{\alpha}}$ by (1.26) and (1.31). We then have

$$\dot{\mathcal{E}}_{\hat{\alpha}}(a(1 \otimes x)) = a\dot{\mathcal{E}}_{\hat{\alpha}}(1 \otimes x) = \psi_G(x)a \ .$$

Thus, choosing $x \in \mathfrak{m}_{\psi_G}$ so that $\psi_G(x) \neq 0$, we conclude $a \in \dot{\mathcal{E}}_{\hat{\alpha}}(\mathfrak{p}_{\hat{\alpha}})$. Hence $(\mathfrak{M} \times_\alpha G)^{\hat{\alpha}} \subset \dot{\mathcal{E}}_{\hat{\alpha}}(\mathfrak{p}_{\hat{\alpha}})$.

(1.33): Recalling (I.2.11), we have for each $\omega \in \mathfrak{n}_*$ and $f \in L^2(G)$,

$$\begin{aligned}
&\langle \hat{\delta}(1 \otimes g), \omega \otimes \omega_f \otimes \Delta \rangle \\
&= \langle 1 \otimes \lambda(g), \omega \otimes \omega_f \otimes \Delta \rangle \\
&= \langle 1,\omega \rangle \int_G g(t^{-1}s)|f(s)|^2 \Delta(t)\, ds\, dt \\
&= \langle 1,\omega \rangle\, \mu_G(g) \langle 1,\omega_f \rangle \\
&= \mu_G(g) \langle 1, \omega \otimes \omega_f \rangle \ .
\end{aligned}$$

Thus (1.2) yields the conclusion.

(1.34): The proof is similar to that for (1.32).

(1.35): For a $y \in \mathfrak{n}$, $\omega \in \mathfrak{n}_*$ and $h,k \in L^2(G)$ we compute:

$$\begin{aligned}
&\langle \dot{\mathcal{E}}_{\hat{\delta}}((1 \otimes g^*)\delta(y)(1 \otimes g)), \omega \otimes \omega_{h,k} \rangle \\
&= \int_G \langle \hat{\delta}_s((1 \otimes g^*)\delta(y)(1 \otimes g)), \omega \otimes \omega_{h,k} \rangle \Delta(s) ds \\
&= \int_G \langle (1 \otimes \lambda_s(g^*))\delta(y)(1 \otimes \lambda_s(g)), \omega \otimes \omega_{h,k} \rangle \Delta(s) ds \\
&= \int_G \left\langle \delta(y), \omega \otimes \omega_{\lambda_s(g)h, \lambda_s(g)k} \right\rangle \Delta(s) ds \ .
\end{aligned}$$

Recalling that $\delta(y) \in \mathfrak{n} \,\overline{\otimes}\, \mathcal{R}(G)$, we compute the following in $\mathcal{R}(G)_* = A(G)$:

$$\left\langle \rho(r), \int_G \omega_{\lambda_s(g)h,\lambda_s(g)k} \Delta(s)ds \right\rangle$$

$$= \int_G (\rho(r)\lambda_s(g)h | \lambda_s(g)k)\Delta(s)ds$$

$$= \iint_{G\times G} g(s^{-1}tr)h(tr)\overline{g(s^{-1}t)k(t)}\Delta(s)dt\, ds$$

$$= \iint_{G\times G} g(str)h(tr)\overline{g(st)k(t)}ds\, dt$$

$$= \int_{G\times G}\left(\int \overline{g^{\vee}(rs^{-1})}g(s)ds\right) h(tr)\overline{k(t)}dt$$

$$= \int_G (g^{\#} * g)(r)h(tr)\overline{k(t)}dt = \varphi(r)(\rho(r)h|k) .$$

Thus, we get

$$\int_G \omega_{\lambda_s(g)h,\lambda_s(g)k} \Delta(s)ds = \varphi\omega_{h,k} .$$

Therefore, we finally obtain the following:

$$\langle \dot{\mathcal{E}}_{\delta}((1 \otimes g^*)\delta(y)(1 \otimes g)), \omega \otimes \omega_{h,k}\rangle$$

$$= \langle \delta(y), \omega \otimes \varphi\omega_{h,k}\rangle$$

$$= \langle \delta \circ \delta_{\varphi}(y), \omega \otimes \omega_{h,k}\rangle \qquad \text{Q.E.D.}$$

Let $\{\pi_\varphi, \mathfrak{H}_\varphi, \eta_\varphi\}$ and $\{\pi_\psi, \mathfrak{H}_\psi, \eta_\psi\}$ be the GNS representation of $\mathfrak{M}$ with respect to faithful semi-finite normal weights φ and ψ on $\mathfrak{M}$. Let $\mathfrak{n}_\varphi$ (resp. $\mathfrak{n}_\psi$) be the set of $x \in \mathfrak{M}$ with $\varphi(x^*x) < \infty$ (resp. $\psi(x^*x) < \infty$). We denote by $S_{\psi,\varphi}$ the closure of the conjugate linear map: $\eta_\varphi(x) \mapsto \eta_\psi(x^*)$, $x \in \mathfrak{n}_\varphi \cap \mathfrak{n}_\psi^*$. Denote the polar decomposition by

$$S_{\psi,\varphi} = J_{\psi,\varphi}\Delta_{\psi,\varphi}^{1/2} .$$

Then $J_{\psi,\varphi}$ is a conjugate linear isometry of $\mathfrak{H}_\varphi$ onto $\mathfrak{H}_\psi$ and $\Delta_{\psi,\varphi}$ is a non-singular positive self-adjoint operator on $\mathfrak{H}_\varphi$.

<u>Proposition</u> 1.7. Let φ be a faithful semi-finite normal weight on $\mathfrak{M}$ and $\tilde{\varphi} = \varphi \circ \alpha^{-1} \circ \mathcal{E}_{\mathfrak{A}}$. Then

(i) $\tilde{\varphi}$ is a faithful semi-finite normal weight on $\mathfrak{M} \times_\alpha G$;

(ii) if J_φ and $J_{\tilde{\varphi}}$ are the modular unitary involutions and u is the unitary representation of G canonically implementing α, then $J_{\tilde{\varphi}} = (J_\varphi \otimes J)u^* = u\,(J_\varphi \otimes J)$.

Proof. Assume that $\mathfrak{M}$ is standard, i.e., $\{\mathfrak{M},\mathfrak{H}\} = \{\pi_\varphi(\mathfrak{M}),\mathfrak{H}_\varphi\}$. We set $\mathfrak{n} = \mathcal{K}(G,\mathfrak{M})\mathfrak{n}_\varphi$.[6] Then $\mathfrak{n} \cap \mathfrak{n}^*$ turns out to be an involutive algebra with respect to

$$(x * y)(s) = \int x(r)\alpha_r(y(r^{-1}s))\,dr$$

$$x^\#(s) = \Delta(s)\alpha_s\,(x(s^{-1}))^* .$$

We denote the involution preserving isomorphism of $\mathfrak{n} \cap \mathfrak{n}^\#$ into $\mathfrak{M} \times_\alpha G$ by π:

$$\pi(x) = \int \alpha(x(s))(1 \otimes \rho(s))\,ds , \qquad x \in \mathfrak{n} \cap \mathfrak{n}^\#.$$

Then $\pi(\mathfrak{n} \cap \mathfrak{n}^\#)$ is σ-weakly dense in $\mathfrak{M} \times_\alpha G$. We denote the linear map of $\mathfrak{n}$ into $\mathfrak{H} \otimes L^2(G)$ by η:

$$(\eta(x))(s) = \eta_\varphi(x(s)), \qquad x \in \mathfrak{n}, \quad s \in G .$$

Then $\eta(\mathfrak{n} \cap \mathfrak{n}^\#)$ is a left Hilbert algebra with respect to

$$\eta(x)\eta(y) = \eta(x * y), \qquad \eta(x)^\# = \eta(x^\#)$$

and dense in $\mathfrak{H} \otimes L^2(G)$. Define $\tilde{J}$ and $\tilde{\Delta}$ by

$$\tilde{J} = (J_\varphi \otimes J)u^* = u\,(J_\varphi \otimes J) ,$$

$$(\tilde{\Delta}^{it}\xi)(s) = \Delta(s)^{-it}\Delta^{it}_{\varphi\circ\alpha_s^{-1},\varphi}\,\xi(s), \qquad \xi \in \mathfrak{H} \otimes L^2(G), \quad t \in \mathbb{R} .$$

Since $J_{\psi,\varphi} = J_\varphi$ by the uniqueness of standard form and so, for $x \in \mathfrak{n} \cap \mathfrak{n}^\#$,

$$\begin{aligned}(\tilde{J}\eta(x^\#))(s) &= \Delta(s)^{1/2}u(s)\ J_\varphi\eta_\varphi(x^\#(s^{-1})) \\ &= \Delta(s)^{-1/2}u(s)\ J_\varphi\eta_\varphi(\alpha_s^{-1}(x(s)^*)) \\ &= \Delta(s)^{-1/2}J_\varphi\eta_{\varphi\circ\alpha_s^{-1}}(x(s)^*) \\ &= \Delta(s)^{-1/2}J_\varphi S_{\varphi\circ\alpha_s^{-1},\varphi}\eta_\varphi(x(s)) \\ &= (\tilde{\Delta}^{1/2}\eta(x))(s) ,\end{aligned}$$

it follows that $\tilde{J}\tilde{\Delta}^{1/2}$ is the closure of the involution $\#$ of the left Hilbert algebra $\eta(\mathfrak{n} \cap \mathfrak{n}^\#)$. The canonical weight ψ on $\mathfrak{M} \times_\alpha G$ is given by

---

6) $\mathcal{K}(G,\mathfrak{M})$ = The space of $\mathfrak{M}$-valued continuous functions on G with compact support.

$$\psi(\pi(x)^*\pi(x)) = (\eta(x)|\eta(x)) = \int \varphi(x(t)^*x(t))\,dt$$

$$= \varphi((x^\# * x)(e)) = \varphi \circ \alpha^{-1} \circ \mathcal{E}_{\alpha}(\pi(x)^*\pi(x)) \,.$$

Therefore, $S_\psi = \tilde{J}\tilde{\Delta}^{1/2}$ and hence $J_\psi = \tilde{J}$ or $J_{\tilde{\varphi}} = \tilde{J}$. Q.E.D.

NOTES

The general theory of operator valued weight was developed by Haagerup, [33]. On the other hand, operator valued weights associated with an action or a co-action are treated independently in [14,34,42,46,60].

§2. Integrability and operator w.

We begin with the definition of integrability:

Definition 2.1. An action α (resp. co-action δ) of G on $\mathfrak{M}$ (resp. $\mathfrak{N}$) is said to be integrable, if the operator valued weight $\mathcal{E}_\alpha$ (resp. $\mathcal{E}_\delta$) defined in §1 is semi-finite, namely, $\mathfrak{p}_\alpha$ (resp. $\mathfrak{p}_\delta$) is σ-weakly dense in $\mathfrak{M}$ (resp. $\mathfrak{N}$).

For example, α_G and δ_G are both integrable. By virtue of Proposition 1.6 a dual co-action and a dual action are integrable.

Lemma 2.2. (a) If α is integrable and $\omega = \omega_o \circ \mathcal{E}_\alpha$ (or $(\omega_o \otimes \mu_G) \circ \alpha$), with any $\omega_o \in \mathfrak{M}_*^+$, then we have

(2.1) $$(\omega \otimes g)(\alpha(x)) = \omega(x)\mu_G(g), \quad x \in \mathfrak{p}_\alpha, \quad g \in L^1(G)_+ .$$

(b) If δ is integrable and $\omega = \omega_o \circ \mathcal{E}_\delta$ (or $(\omega_o \otimes \psi_G) \circ \delta$), with any $\omega_o \in \mathfrak{N}_*^+$, then

(2.2) $$(\omega \otimes \varphi)(\delta(y)) = \omega(y)\, \varphi(1), \quad y \in \mathfrak{p}_\delta, \quad \varphi \in A(G)_+ .$$

Proof. (a) Our assertion follows from the fact that:

$$\mathcal{E}_\alpha(\alpha_s(x)) = \Delta(s)^{-1}\mathcal{E}_\alpha(x) \quad \text{for} \quad x \in \mathfrak{p}_\alpha .$$

(b) We may assume that $y \in \mathfrak{p}_\delta^+$. Then we have

$$\begin{aligned}
(\omega \otimes \varphi)(\delta(y)) &= \omega(\delta_\varphi(y)) = \langle \mathcal{E}_\delta(\delta_\varphi(y)), \omega_o \rangle \\
&= \sup_{\psi \in F} \langle \delta_\psi(\delta_\varphi(y)), \omega_o \rangle = \sup_{\psi \in F} \langle \delta_\varphi(\delta_\psi(y)), \omega_o \rangle \\
&= \langle \delta(\mathcal{E}_\delta(y)), \omega_o \otimes \varphi \rangle \\
&= \langle \mathcal{E}_\delta(y) \otimes 1, \omega_o \otimes \varphi \rangle \quad \text{by} \quad (1.24), \\
&= (\omega \otimes \varphi)(y \otimes 1) .
\end{aligned}$$

Q.E.D.

We now present the main result of this section:

Theorem 2.3. If $\mathfrak{N}$ is standard and δ is integrable on $\mathfrak{N}$, then there exists a unitary w in $\mathcal{L}(\mathfrak{K}) \overline{\otimes} \mathcal{R}(G)$ such that

(2.3) $$\delta(z) = w^*(z \otimes 1)w ,$$

(2.4) $$(w^* \otimes 1)(\iota \otimes \sigma)(w^* \otimes 1) = (\iota \otimes \delta_G)(w^*) .$$

Proof. Let $\omega = \omega_o \circ \mathcal{E}_\delta$ for a faithful normal state ω_o on $\mathfrak{N}$. Since δ is integrable, $\omega \otimes \psi_G$ is a faithful semi-finite, normal weight on $\mathfrak{N} \overline{\otimes} \mathcal{R}(G)$. Since $\mathfrak{N} \overline{\otimes} \mathcal{R}(G)$ is standard, it is assumed to act on the L^2-completion of $\mathfrak{n}_\omega \otimes \mathfrak{n}_{\psi_G}$.

For any $x_j \in n_\omega$ and $y_j \in n_{\psi_G}$, we have

$$\|\sum x_j \otimes y_j\|^2_{\omega\otimes\psi_G} = (\omega \otimes \psi_G)(\sum x_k^* x_j \otimes y_k^* y_j)$$
$$= \sum (\omega \otimes y_j \psi_G y_k^*)(x_k^* x_j \otimes 1)$$
$$= \sum (\omega \otimes y_j \psi_G y_k^*)(\delta(x_k^* x_j)) \quad \text{by (2.2)},$$
$$= (\omega \otimes \psi_G)(\sum (1 \otimes y_k)^* \delta(x_k^* x_j)(1 \otimes y_j))$$
$$= \|\sum \delta(x_j)(1 \otimes y_j)\|^2_{\omega\otimes\psi_G} .$$

Therefore we have

$$\sum_{j=1}^{n} \delta(x_j)(1 \otimes y_j) \in n_{\omega\otimes\psi_G} \quad \text{for } x_j \in n_\omega , \quad y_j \in n_{\psi_G} \quad (j=1,\dots,n) .$$

We define an isometry w^* on $L^2(\mathfrak{n} \,\bar{\otimes}\, \mathcal{R}(G), \omega \otimes \psi_G) = \mathfrak{K} \otimes L^2(G)$ by

$$(2.6) \qquad w^* \eta_{\omega\otimes\psi_G}(x \otimes y) = \eta_{\omega\otimes\psi_G}(\delta(x)(1 \otimes y)) .$$

For the moment we assume that w is unitary, which will be proved later in a series of lemmas.

For any $z \in \mathfrak{n}$, we have

$$w^*(z \otimes 1) w \eta_{\omega\otimes\psi_G}(\delta(x)(1 \otimes y))$$
$$= w^*(z \otimes 1)\eta_{\omega\otimes\psi_G}(x \otimes y)$$
$$= w^* \eta_{\omega\otimes\psi_G}(zx \otimes y)$$
$$= \eta_{\omega\otimes\psi_G}(\delta(zx)(1 \otimes y))$$
$$= \delta(z)\eta_{\omega\otimes\psi_G}(\delta(x)(1 \otimes y)) .$$

By the assumption on w, the set of all $\delta(x)(1 \otimes y)$, $x \in n_\omega$, $y \in n_{\psi_G}$ is total in $\mathfrak{K} \otimes L^2(G)$. Thus (2.3) is obtained.

Now, if $f, g \in L^1(G) \cap L^2(G)$, then

$$\delta_G(\rho(f^\flat))(1 \otimes \rho(g^\flat)) = (\rho \otimes \rho)((W_G^*(f \otimes g))^{\flat\otimes\flat})$$

and so

$$W_G^* \eta_{\psi_G\otimes\psi_G}(\rho(f^\flat) \otimes \rho(g^\flat)) = W_G^*(\bar{f} \otimes \bar{g})$$
$$(2.7) \qquad = \eta_{\psi_G\otimes\psi_G}((\rho \otimes \rho)((W_G^*(f \otimes g))^{\flat\otimes\flat})$$
$$= \eta_{\psi_G\otimes\psi_G}(\delta_G(\rho(f^\flat))(1 \otimes \rho(g^\flat))) .$$

Therefore, if $x \in \mathfrak{n}_\omega$ and $a,b \in \mathfrak{n}_{\psi_G}$, then with $\Psi = \omega \otimes \psi_G \otimes \psi_G$ we have

$$(w^* \otimes 1)(\iota \otimes \sigma)(w^* \otimes 1)\eta_\Psi((x \otimes \delta_G(a))(1 \otimes 1 \otimes b))$$

$$= (w^* \otimes 1)\eta_\Psi(\{(\iota \otimes \sigma) \circ (\delta \otimes \iota)(x \otimes 1)\}(1 \otimes \delta_G(a))(1 \otimes 1 \otimes b))$$

$$= \eta_\Psi(((\delta \otimes \iota)\delta(x))(1 \otimes \delta_G(a))(1 \otimes 1 \otimes b))$$

$$= \eta_\Psi((\iota \otimes \delta_G)(\delta(x)(1 \otimes a))(1 \otimes 1 \otimes b))$$

$$= (1 \otimes W_G^*)\eta_\Psi((\delta \otimes \iota)(x \otimes 1)(1 \otimes a \otimes b)) \quad \text{by (2.7)},$$

$$= (1 \otimes W_G^*)(w^* \otimes 1)\eta_\Psi(x \otimes a \otimes b)$$

$$= (1 \otimes W_G^*)(w^* \otimes 1)(1 \otimes W_G)\eta_\Psi((x \otimes \delta_G(a))(1 \otimes 1 \otimes b)),$$

which entails (2.4) Q.E.D.

Before going into our lemmas, we note the following facts:

(2.8) $$\delta_G(\mathfrak{n}_{\psi_G}) \subset \mathfrak{n}_{\psi_G \otimes \psi_G};$$

(2.9) $$\delta_G(\mathfrak{n}_{\psi_G})(1 \otimes \mathfrak{n}_{\psi_G}) \subset \mathfrak{n}_{\psi_G \otimes \psi_G};$$

(2.10) $$\psi_G(\rho(f)y\rho(g)^*) = \psi_G(y\rho(g^\# * f)).$$

Indeed, (2.8) and (2.9) were proved in the above proof. Equality (2.10) follows from the calculation:

(2.11) $$\begin{aligned}\psi_G(\rho(f)\rho(s)\rho(g)^*) &= (f|\rho(s)g) = (g^\# * f)(s^{-1})\\ &= (\rho(s)^*(g^\# * f))(e) = \psi_G(\rho(s)\rho(g^\# * f)).\end{aligned}$$

<u>Lemma</u> 2.4. If $\omega = \omega_o \circ \mathcal{E}_\delta$ for some faithful $\omega_o \in \mathfrak{h}_*^+$, then

(2.12) $$(\omega \otimes \psi_G)(\delta(a)^*(b \otimes \rho(h))) = (\omega \otimes \psi_G)((a^* \otimes 1)\delta(b)(1 \otimes \rho(h^\vee)))$$

for all $a,b \in \mathfrak{n}_\omega$ and $h \in A(G) \cap \mathcal{K}(G)$.

<u>Proof</u>. For any $f,g,h \in A(G) \cap \mathcal{K}(G)$, we have

(2.13) $$(\psi_G \otimes \psi_G)(\delta_G(\rho(f))(\rho(g) \otimes \rho(h))) = (\psi_G \otimes \psi_G)((\rho(f) \otimes 1)\,\delta_G(\rho(g))(1 \otimes \rho(h^\vee)))$$

by direct computation. Since $h \in A(G)$, it is of the form $h_2^\# * h_1$ for some $h_1,h_2 \in L^2(G)$. Put $\Psi = \omega_o \otimes \psi_G \otimes \psi_G$. For each $x,y \in \mathfrak{h}$, we have, by (2.10),

$$
\begin{aligned}
&(y \otimes \rho(g) \otimes \rho(h_2)^* | (\iota \otimes \delta_G)(x \otimes \rho(f))^*(1 \otimes 1 \otimes \rho(h_1)^*))_\Psi \\
&\quad = \Psi((\iota \otimes \delta_G)(x \otimes \rho(f))(y \otimes \rho(g) \otimes \rho(h_2^\# * h_1))) \\
(2.14) \qquad &\quad = \Psi((x \otimes \rho(f) \otimes 1)(\iota \otimes \delta_G)(y \otimes \rho(g))(1 \otimes 1 \otimes \rho(h_1^\vee * \bar{h}_2))) \\
&\quad = ((\iota \otimes \delta_G)(y \otimes \rho(g))(1 \otimes 1 \otimes \rho(\bar{h}_1)^*) | (x \otimes \rho(f) \otimes \rho(\bar{h}_2))^*)_\Psi ,
\end{aligned}
$$

where the second equality follows from (2.13) and

$$
(2.15) \qquad (h_2^\# * h_1)^\vee = h_1^\vee * \bar{h}_2 \ .
$$

Since $K(G) \subset L^2(G)$, it follows that

$$
y \otimes \rho(g) \otimes \rho(h_2)^* \in \mathfrak{n}_\Psi \quad \text{and} \quad \rho(h_1)^* , \rho(\bar{h}_2)^* \in \mathfrak{n}_{\psi_G} .
$$

Since $\delta_G(\rho(g))(1 \otimes \rho(\bar{h}_1)^*) \in \mathfrak{n}_{\psi_G \otimes \psi_G}$, it follows that

$$
(\iota \otimes \delta_G)(y \otimes \rho(g))(1 \otimes 1 \otimes \rho(\bar{h}_1)^*) \in \mathfrak{n}_\Psi .
$$

We fix these elements. Then (2.14) is extended uniquely to $L^2(\mathcal{h} \bar{\otimes} \mathcal{R}(G), \omega_o \otimes \psi_G)$ as a bounded linear functional of $x \otimes \rho(f)$. Since $a \in \mathfrak{n}_\delta$, we have $\delta(a) \in \mathfrak{n}_{\omega_o \otimes \psi_G}$. Replacing $(x \otimes \rho(f))^*$ by $\delta(a)$, we have

$$
\begin{aligned}
&(y \otimes \rho(g) \otimes \rho(h_2)^* | (\delta \otimes \iota)(\delta(a))(1 \otimes 1 \otimes \rho(h_1)^*))_\Psi \\
(2.16) \qquad &\quad = ((\iota \otimes \delta_G)(y \otimes \rho(g))(1 \otimes 1 \otimes \rho(\bar{h}_1)^*) | \delta(a) \otimes \rho(\bar{h}_2)^*)_\Psi .
\end{aligned}
$$

Similarly, $\rho(h_2)^*$, $\rho(\bar{h}_1)^* \in \mathfrak{n}_{\psi_G}$ and $\delta(a) \otimes \rho(\bar{h}_2)^* \in \mathfrak{n}_\Psi$. Since $\delta(a)(1 \otimes \rho(h_1)^*) \in \mathfrak{q}_{\omega_o \otimes \psi_G}$, it follows that

$$
(\delta \otimes \iota)(\delta(a))(1 \otimes 1 \otimes \rho(h_1)^*) \in \mathfrak{n}_\Psi .
$$

Therefore, (2.16) is extended uniquely to $L^2(\mathcal{h} \bar{\otimes} \mathcal{R}(G), \omega_o \otimes \psi_G)$ as a bounded linear functional of $y \otimes \rho(g)$. Since $b \in \mathfrak{q}_\delta$, we have $\delta(b) \in \mathfrak{n}_{\omega_o \otimes \psi_G}$. Replacing $y \otimes \rho(g)$ by $\delta(b)$, we have

$$
\begin{aligned}
&(\delta(b) \otimes \rho(h_2)^* | (\delta \otimes \iota)(\delta(a))(1 \otimes 1 \otimes \rho(h_1)^*))_\Psi \\
(2.17) \qquad &\quad = ((\delta \otimes \iota)(\delta(b))(1 \otimes 1 \otimes \rho(\bar{h}_1)^*) | \delta(a) \otimes \rho(\bar{h}_2)^*)_\Psi .
\end{aligned}
$$

Since $a,b \in \mathfrak{q}_\delta$, each argument belongs to $\mathfrak{n}_\Psi$. Therefore, (2.17) gives our desired result (2.12) by using (2.10) and (2.15). Q.E.D.

<u>Lemma</u> 2.5. If ω is a faithful, semi-finite, normal weight on $\mathcal{h}$ satisfying (2.12), then

$$
(2.18) \qquad \delta \circ \sigma_t^\omega = (\sigma_t^\omega \otimes \iota) \circ \delta .
$$

<u>Proof</u>. By virtue of the KMS condition, for any $x, y \in \mathfrak{n}_{\omega \otimes \psi_G}$ and $f \in A(G) \cap (G)$, there exists a function F continuous on and holomorphic in the strip $\{\mu \in \mathbf{C} : 0 \leq \operatorname{Im} \mu \leq 1\}$ such that

$$F(t) = (\omega \otimes \psi_G)(\sigma_t^{\omega \otimes \psi_G}(x \otimes \rho(\Delta^{-1} f^{\vee}))\, \delta(y)) \;;$$

$$F(t + i) = (\omega \otimes \psi_G)(\delta(y)\sigma_t^{\omega \otimes \psi_G}(x \otimes \rho(\Delta^{-1} f^{\vee}))) \;.$$

If a is an analytic element in $\mathcal{R}(G)$ with respect to σ^{ψ_G}, then

$$(2.19) \qquad \psi_G(ab) = \psi_G(b\sigma_{-i}^{\psi_G}(a)) \;.$$

Since

$$(2.20) \qquad \sigma_t^{\psi_G}(\rho(f)) = \rho(\Delta^{it}\; f) \;, \qquad t \in \mathbb{R} \;,$$

it follows that

$$\begin{aligned} F(t) &= (\omega \otimes \psi_G)((\sigma_t^{\omega}(x) \otimes 1)\delta(y)(1 \otimes \sigma_t^{\psi_G}(\rho(f^{\vee})))) \\ (2.21) \qquad &= (\omega \otimes \psi_G)(\delta(\sigma_t^{\omega}(x))(y \otimes \sigma_{-t}^{\psi_G}(\rho(f)))) \qquad \text{by } (2.12) \\ &= (\omega \otimes \psi_G)((\iota \otimes \sigma_t^{\psi_G})(\delta(\sigma_t^{\omega}(x))\; (y \otimes \rho(f)))) \;, \end{aligned}$$

and

$$\begin{aligned} F(t + i) &= (\omega \otimes \psi_G)((y \otimes 1)\delta(\sigma_t^{\omega}(x))(1 \otimes \sigma_{-t}^{\psi_G}(\rho(\Delta\;\; f)))) \quad \text{by } (2.12) \\ (2.22) \qquad &= (\omega \otimes \psi_G)((y \otimes \sigma_{-t}^{\psi_G}(\rho(f)))\delta(\sigma_t^{\omega}(x))) \\ &= (\omega \otimes \psi_G)((y \otimes \rho(f))(\iota \otimes \sigma_t^{\psi_G})(\delta(\sigma_t^{\omega}(x)))) \;. \end{aligned}$$

If $z \in \mathfrak{n}_{\omega}$, then $\delta(z)$ is approximated by a sequence $\{z_n\}$ in $\mathfrak{n}_{\omega} \otimes \mathfrak{n}_{\psi_G}$ such that $\|z_n - \delta(z)\|_{\omega \otimes \psi_G}$ and $\|z_n^* - \delta(z)^*\|_{\omega \otimes \psi_G}$ converge to 0. Therefore, by (2.21) and (2.22), there exists a sequence of functions G_n continuous on and holomorphic in the strip: $0 \leq \operatorname{Im} \mu \leq 1$ such that

$$G_n(t) = (\omega \otimes \psi_G)((\iota \otimes \sigma_t^{\psi_G})(\delta(\sigma_t^{\omega}(x)))z_n)$$

$$G_n(t + i) = (\omega \otimes \psi_G)(z_n(\iota \otimes \sigma_t^{\psi_G})(\delta(\sigma_t^{\omega}(x)))) \;.$$

Further we set

$$(2.23) \qquad \begin{cases} G(t) = (\omega \otimes \psi_G)((\iota \otimes \sigma_t^{\psi_G})(\delta(\sigma_t^{\omega}(x)))\delta(z)) \\ G\;(t + i) = (\omega \otimes \psi_G)(\delta(z)(\iota \otimes \sigma_t^{\psi_G})(\delta(\sigma_t^{\omega}(x)))) \;. \end{cases}$$

Since

$$|G_n(t) - G(t)| \leq \|z_n - \delta(z)\|_{\omega\otimes\psi_G} \|\delta(\sigma_t^\omega(x))\|_{\omega\otimes\psi_G}$$

$$|G_n(t+i) - G(t+i)| \leq \|z_n^* - \delta(z)^*\|_{\omega\otimes\psi} \|\delta(\sigma_t^\omega(x))^*\|_{\omega\otimes\psi_G} ,$$

it follows that $G_n(t)$ (resp. $G_n(t+i)$) converges uniformly to $G(t)$ (resp. $G(t+i)$) on each compact subset of $\mathbb{R}$. The three lines theorem implies that there is a holomorphic extension of G to the strip $0 \leq \operatorname{Im}\mu \leq 1$, denoted by the same symbol G, such that G_n converges uniformly to G on each compact subset of the same strip.

Finally we shall go into the proof of (2.18). As $\delta(\mathfrak{n})$ is globally invariant under $\iota\otimes\sigma_t^{\psi_G}$ by (2.20), there is an action γ of $\mathbb{R}$ on $\mathfrak{n}$ such that

$$(\iota\otimes\sigma_t^{\psi_G})(\delta(y)) = \delta(\gamma_t(y)) , \qquad y\in\mathfrak{n} .$$

Since $\omega = (\omega_0\otimes\psi_G)\circ\delta$, ω is γ invariant. Hence γ commutes with σ^ω. We define σ_t on $\delta(\mathfrak{n})$ by

$$\sigma_t(\delta(y)) = (\iota\otimes\sigma_t^{\psi_G})(\delta(\sigma_t^\omega(y))) .$$

Then σ is an action of $\mathbb{R}$ on $\delta(\mathfrak{n})$. Indeed,

$$\sigma_{s+t}(\delta(y)) = (\iota\otimes\sigma_{s+t}^{\psi_G})\circ\delta\circ\sigma_{s+t}^\omega(y) = (1\otimes\sigma_s^{\psi_G})\circ\delta\circ\sigma_s^\omega\circ(\gamma_t\circ\sigma_t^\omega(y))$$

$$= \sigma_s(\delta(\gamma_t\circ\sigma_t^\omega(y))) = \sigma_s(\sigma_t(\delta(y)))$$

and the map: $t\mapsto\sigma_t(\delta(y))$ is σ-weakly continuous for each $y\in\mathfrak{n}$. Therefore, (2.23) indicates that σ is the KMS automorphism of $\omega\otimes\psi_G\restriction\delta(\mathfrak{n})$, i.e. $\sigma_t = \sigma_t^\omega\otimes\sigma_t^{\psi_G}$ on $\delta(\mathfrak{n})$. Consequently, $(\iota\otimes\sigma_t^{\psi_G})\circ\delta\circ\sigma_t^\omega = (\sigma_t^\omega\otimes\sigma_t^{\psi_G})\circ\delta$ on $\mathfrak{n}$ and (2.18) holds on $\mathfrak{n}$.
Q.E.D.

The following lemma verifies the assumption on w in the proof of Theorem 2.3:

Lemma 2.6. If w is a coisometry defined by (2.6), then w is a unitary satisfying

(2.24) $$w(\Delta_\omega\otimes 1)w^* = \Delta_\omega\otimes 1 ;$$

(2.25) $$w^*(J_\omega\otimes C) = (J_\omega\otimes C)w ,$$

where Δ_ω and J_ω are the modular operator and the modular unitary involution for ω, and C is the conjugation: $f\in L^2(G)\to\bar{f}\in L^2(G)$.

Proof. Let $\mathfrak{q}_\omega^a$ be the set of analytic elements in $\mathfrak{n}$ with respect to σ^ω. If $z\in\mathfrak{q}_\omega^a$, then

$$(\Delta_\omega^{\frac{1}{2}} \otimes 1)w^*(\Delta_\omega^{-\frac{1}{2}} \otimes 1)\eta_{\omega\otimes\psi_G}(z \otimes \rho(h))$$

$$= (\Delta_\omega^{\frac{1}{2}} \otimes 1)\eta_{\omega\otimes\psi_G}(\delta(\sigma^\omega_{i/2}(z))(1 \otimes \rho(h)))$$

(2.26) $$= (\Delta_\omega^{\frac{1}{2}} \otimes 1)\eta_{\omega\otimes\psi_G}((\sigma^\omega_{i/2} \otimes \iota)(\delta(z)(1 \otimes \rho(h)))) \text{ by (2.18)},$$

$$= \eta_{\omega\otimes\psi_G}(\delta(z)(1 \otimes \rho(h)))$$

$$= w^*\eta_{\omega\otimes\psi_G}(z \otimes \rho(h)) ,$$

which gives $(\Delta_\omega^{\frac{1}{2}} \otimes 1)w^*(\Delta_\omega^{-\frac{1}{2}} \otimes 1) = w^*$ on $q_\omega^a \otimes n_{\psi_G}$.

For any $x,y \in q_\omega^a$ and $f,g \in \mathcal{K}(G)$ we have

$$(w^*(x \otimes \rho(f)^*) | y \otimes \rho(g)^*)_{\omega\otimes\psi_G}$$

$$= (\omega \otimes \psi_G)((y^* \otimes 1)\delta(x)(1 \otimes \rho(f^\# * g)))$$

$$= (\omega \otimes \psi_G)(\delta(y^*)(x \otimes \rho(g^\vee * \bar{f}))) \qquad \text{by (2.12)},$$

$$= (\omega \otimes \psi_G)(\delta(y^*)(1 \otimes \rho(g^\vee))(x \otimes \rho(\bar{f})))$$

$$= (\omega \otimes \psi_G)((\sigma^\omega_i(x) \otimes \rho(\Delta\bar{f}))\delta(y^*)(1 \otimes \rho(g^\vee))) \qquad \text{by (2.19)},$$

$$= (w^*(y^* \otimes \rho(g^\vee)) | \sigma^\omega_{-i}(x^*) \otimes \rho(f^\vee))_{\omega\otimes\psi_G}$$

$$= (w^*(\Delta_\omega^{-\frac{1}{2}} J_\omega y \otimes C\rho(g)^*) | \Delta_\omega^{\frac{1}{2}} J_\omega x \otimes C\rho(f)^*)_{\omega\otimes\psi_G}$$

$$= ((\Delta_\omega^{\frac{1}{2}} \otimes 1)w^*(\Delta_\omega^{-\frac{1}{2}} \otimes 1)(J_\omega \otimes C)(y \otimes \rho(g)^*) | (J_\omega \otimes C)(x \otimes \rho(f)^*))_{\omega\otimes\psi_G}$$

$$= (x \otimes \rho(f)^* | (J_\omega \otimes C)w^*(J_\omega \otimes C)(y \otimes \rho(g)^*))_{\omega\otimes\psi_G} , \qquad \text{by (2.26)} .$$

Thus w^* satisfies (2.25) and so w is a unitary. Using the same computation as (2.26) we have

$$w(\Delta_\omega^{\frac{1}{2}} \otimes 1)w^* = \Delta_\omega^{\frac{1}{2}} \otimes 1 \text{ on } q_\omega^a \otimes n_{\psi_G} .$$

Since both sides are self-adjoint operators with the same core $q_\omega^a \otimes n_{\psi_G}$, we have (2.24). Q.E.D.

NOTES

The importance of the integrable action was first pointed out in [14]. Lemma 2.5 is taken from [60]. The integrability for the Kac algebra version was given in [25].

§3. Integrable actions and co-actions.

In this section we shall give a characterization of integrable actions and co-actions, assuming the proper infiniteness of the fixed points algebras.

Theorem 3.1. If $\mathfrak{M}^{\alpha}$ is properly infinite, the following three conditions are equivalent:

(i) α is integrable;

(ii) For any non zero projection $f \in \mathfrak{M}^{\alpha} \otimes C$, there exists a non zero $x \in \bar{\mathfrak{M}}$ such that $x = fxf$ and

$$(3.1) \qquad (1 \otimes V_G')\bar{\alpha}(x) = x \otimes 1 ,$$

or equivalently

$$(\alpha_t \otimes \iota)(x) = (1 \otimes \lambda(t))^* x ;$$

(iii) $\{\bar{\mathfrak{M}},\bar{\alpha}\} \cong \{\bar{\mathfrak{M}},\tilde{\alpha}\}_e$ for some projection e in $\bar{\mathfrak{M}}^{\tilde{\alpha}}$.

Proof. (i) $\Rightarrow$ (ii): Suppose α is integrable. Then there exists a non zero $z \in \mathfrak{q}_{\alpha}$ with $z \otimes 1 = f(z \otimes 1)f$. For any $g \in K(G)$ we set

$$(3.2) \qquad (x\xi)(s) = \Delta(s)^{\frac{1}{2}}\alpha_s(z)\eta \ , \quad \eta = \int g(t)\xi(t)dt$$

for $\xi \in \mathfrak{H} \otimes L^2(G)$. Then $fxf = x$. Since $z \neq 0$, we have $x \neq 0$. Since

$$\|x\xi\|^2 = \int \|\alpha_s(z)\eta\|^2 \Delta(s)ds = \langle \mathcal{E}_{\alpha}(z^*z),\omega_{\eta}\rangle$$
$$\leq \|\mathcal{E}_{\alpha}(z^*z)\| \, \|\eta\|^2 \leq \|\mathcal{E}_{\alpha}(z^*z)\| \, \|g\|_2^2 \, \|\xi\|_2^2 ,$$

x is bounded on $\mathfrak{H} \otimes L^2(G)$. If we replace ξ by $(x' \otimes 1)\xi$ with $x' \in \mathfrak{M}'$ in (3.2), then $[x,x' \otimes 1] = 0$ and so $x \in \bar{\mathfrak{M}}$. Since

$$\begin{aligned}
((1 \otimes V_G')\bar{\alpha}(x)\xi)(s,t) &= \Delta(t)^{\frac{1}{2}}(\bar{\alpha}(x)\xi)(t^{-1}s,t) \\
&= \Delta(t)^{\frac{1}{2}}\bar{\alpha}_t(x)\xi(t^{-1}s,t) \\
&= \Delta(t)^{\frac{1}{2}}\Delta(t^{-1}s)^{\frac{1}{2}}\alpha_t(\alpha_{t^{-1}s}(z))\int g(r)\xi(r,t)dr \quad \text{by (3.2)} , \\
&= \Delta(s)^{\frac{1}{2}}\alpha_s(z)\int g(r)\xi(r,t)dr \\
&= ((x \otimes 1)\xi)(s,t) ,
\end{aligned}$$

we have $x = fxf$ satisfying (3.1).

(ii) $\Rightarrow$ (iii): We use the 2×2 matrix method due to Connes. Let $\mathcal{J}$ be the set of all $x \in \bar{\mathfrak{M}}$ satisfying (3.1). Let γ be an action of G on $\bar{\mathfrak{M}} \otimes F_2$ defined by

$$(3.3)\qquad \gamma_t : \begin{pmatrix} x_{11} & x_{12} \\ x_{21} & x_{22} \end{pmatrix} \longrightarrow \begin{pmatrix} \bar{\alpha}_t(x_{11}) & \bar{\alpha}_t(x_{12})(1 \otimes \lambda(t))^* \\ (1 \otimes \lambda(t))\bar{\alpha}_t(x_{21}) & \tilde{\alpha}_t(x_{22}) \end{pmatrix}.$$

Therefore, if $x \in \mathcal{J}$, then $x \otimes e_{21} \in (\bar{\mathfrak{M}} \otimes F_2)^\gamma$. Then condition (ii) implies that the central support of $1 \otimes e_{11}$ in $(\bar{\mathfrak{M}} \otimes F_2)^\gamma$ is majorized by the central support of $1 \otimes e_{22}$ in $(\bar{\mathfrak{M}} \otimes F_2)^\gamma$. Since $\mathfrak{M}^\alpha$ is properly infinite and $\mathfrak{M}^\alpha \otimes \mathbf{C} \otimes e_{22}$ is contained in $((\bar{\mathfrak{M}} \otimes F_2)^\gamma)_{1 \otimes e_{22}}$, $1 \otimes e_{22}$ is also properly infinite in $(\bar{\mathfrak{M}} \otimes F_2)^\gamma$. Moreover, $1 \otimes e_{11}$ is σ-finite in $(\bar{\mathfrak{M}} \otimes F_2)^\gamma$, so $1 \otimes e_{11} < 1 \otimes e_{22}$ in $(\bar{\mathfrak{M}} \otimes F_2)^\gamma$. Therefore, there exists an isometry $w \in \mathcal{J}$, because $w \otimes e_{21} \in (\bar{\mathfrak{M}} \otimes F_2)^\gamma$ implies $w \in \mathcal{J}$ by (3.3). Put $e = ww^*$. Then $e \in \bar{\mathfrak{M}}^{\tilde{\alpha}}$ and (iii) is proved.

(iii) $\Rightarrow$ (i): Since $\tilde{\alpha}$ is integrable on $\bar{\mathfrak{M}}$ and $e \in \bar{\mathfrak{M}}^{\tilde{\alpha}}$, $\tilde{\alpha}^e$ is also integrable on $\bar{\mathfrak{M}}_e$. Thus $\bar{\alpha}$ is integrable on $\bar{\mathfrak{M}}$ by (iii). If p is a minimal projection in $\mathcal{L}(L^2(G))$, then $1 \otimes p \in \bar{\mathfrak{M}}^{\bar{\alpha}}$ and hence $\bar{\alpha}^{1\otimes p}$ is integrable on $\bar{\mathfrak{M}}_{1\otimes p}$. Since $\{\mathfrak{M},\alpha\} = \{\bar{\mathfrak{M}},\bar{\alpha}\}_{1\otimes p}$, α is integrable on $\mathfrak{M}$. Q.E.D.

Next we shall consider the dual version of the above theorem.

<u>Theorem</u> 3.2. If $\mathfrak{N}^\delta$ is properly infinite, the following three conditions are equivalent:

(i) δ is integrable;

(ii) For any non zero projection $f \in \mathfrak{N}^\delta \otimes \mathbf{C}$ there exists a non zero $y \in \bar{\mathfrak{N}}$ such that $y = fyf$ and

$$(3.4)\qquad (1 \otimes W_G)\bar{\delta}(y) = y \otimes 1 .$$

(iii) $\{\bar{\mathfrak{N}},\bar{\delta}\} \cong \{\bar{\mathfrak{N}},\tilde{\delta}\}_e$ for some projection e in $\bar{\mathfrak{N}}^{\tilde{\delta}}$.

<u>Proof</u>. (i) $\Rightarrow$ (ii): Suppose that δ is integrable. There exists a non zero $z \in \mathfrak{q}_\delta$ such that $z = \delta_\rho(z)$ for some $\rho \in A(G) \cap \mathcal{K}(G)$. Fix a non zero $d \in \mathfrak{n}_{\psi_G}$. For any $x_j \in \mathfrak{q}_\delta$ and $a_j \in \mathfrak{n}_{\psi_G}$ we define an operator y by

$$(3.5)\qquad y\eta_{\omega\otimes\psi_G}(\Sigma x_j \otimes a_j) = \Sigma\, \psi_G(d^*a_j)\eta_{\omega\otimes\psi_G}(\delta(z)(x_j \otimes 1)) ,$$

where ω is a faithful, semi-finite, normal weight $(\omega_o \otimes \psi_G)\circ\delta$ on $\mathfrak{N}$ for some faithful $\omega_o \in \mathfrak{N}_*^+$. Since $\langle \mathcal{E}_\delta(z^*z),\varphi\rangle \le \mu_z\|\varphi\| (\varphi \in \mathfrak{N}_*)$ for some $\mu_z > 0$, we have

$$\begin{aligned} &\|\Sigma\, \psi_G(d^*a_j)\delta(z)(x_j \otimes 1)\|^2_{\omega\otimes\psi_G} \\ &= ((\Sigma\, \psi_G(d^*a_k)x_k)^*\omega(\Sigma\, \psi_G(d^*a_j)x_j)\otimes\psi_G)(\delta(z^*z)) \\ &= \langle \mathcal{E}_\delta(z^*z),\ (\Sigma\, \psi_G(d^*a_k)x_k)^*\omega(\Sigma\, \psi_G(d^*a_j)x_j)\rangle \\ &\leqq \mu_z\|\Sigma\, \psi_G(d^*a_j)x_j\|^2_\omega \end{aligned}$$

and hence, by $\|\Sigma\, \psi_G(d^*a_j)x_j\|^2_\omega \le \|d\|^2_{\psi_G}\|\Sigma\, x_j \otimes a_j\|^2_{\omega\otimes\psi_G}$, we have

$$\|\Sigma\, \psi_G(d^*a_j)\delta(z)(x_j \otimes 1)\|^2_{\omega\otimes\psi_G} \le \mu_z\|d\|^2_{\psi_G}\|\Sigma\, x_j \otimes a_j\|^2_{\omega\otimes\psi_G}$$

Since q_δ is dense in $\mathfrak{n}$ and $z \neq 0$, $\delta(z)(x \otimes 1) \neq 0$ for some $x \in q_\delta$. Therefore y is a non zero bounded operator.

In our proof we may assume that $\mathfrak{n}$ is standard. Therefore $\mathfrak{H} \otimes L^2(G)$ is identified with the L^2-completion of $n_{\omega \otimes \psi_G}$, and that $\mathfrak{n}' \otimes C$ acts on it. Since

$$\delta(z)(y'x \otimes 1) = (y' \otimes 1)\delta(z)(x \otimes 1)$$

for all $y' \in \mathfrak{n}'$, (3.5) implies that $[y, y' \otimes 1] = 0$, and hence $y \in \bar{\mathfrak{n}}$.

It remains to show that y satisfies (3.4). Since $z = \delta_\rho(z)$ for $\rho \in A(G) \cap \mathcal{K}(G)$, it follows from Lemma 1.5 that $\delta(z)$ is the σ-weak limit of

$$\int \delta_{\rho(r)*\psi}(z) \otimes \rho(r)dr .$$

For any $b \in n_{\psi_G}$ we set

$$F(r,\psi) = \bar{\delta}((\delta_{\rho(r)*\psi}(z)x) \otimes 1)(1 \otimes \rho(r) \otimes b) . \tag{3.6}$$

Then

$$\begin{aligned} F(r,\psi) &= \bar{\delta}(\delta_{\rho(r)*\psi}(z) \otimes 1)\bar{\delta}(1 \otimes \rho(r))\bar{\delta}(x \otimes 1)(1 \otimes 1 \otimes b) \\ &= \bar{\delta}(\delta_{\rho(r)*\psi}(z) \otimes \rho(r))\bar{\delta}(x \otimes 1)(1 \otimes 1 \otimes b) , \end{aligned}$$

so that

$$\begin{aligned} \lim_\psi \int F(r,\psi)dr &= \bar{\delta}(\delta(z))\bar{\delta}(x \otimes 1)(1 \otimes 1 \otimes b) \\ &= (\iota \otimes \delta_G)(\delta(z))\bar{\delta}(x \otimes 1)(1 \otimes 1 \otimes b) , \end{aligned} \tag{3.7}$$

where the last equality follows from $\sigma \circ \delta_G = \delta_G$. Since $\mathfrak{n}$ is assumed to be standard and δ is integrable, δ is implemented by a unitary w in $\mathcal{L}(\mathfrak{H}) \bar{\otimes} \mathcal{R}(G)$ satisfying (2.3) and (2.4) by Theorem 2.3. Then

$$\bar{\delta}(y) = Ad_{(\iota \otimes \sigma)(w^* \otimes 1)}(y \otimes 1) . \tag{3.8}$$

Put $\Psi = \omega \otimes \psi_G \otimes \psi_G$. Since, for any $x \in q_\delta$ and $a, b \in n_{\psi_G}$, we have

$$((\iota \otimes \sigma)(w^* \otimes 1))(y \otimes 1)((\iota \otimes \sigma)(w \otimes 1))\eta_\Psi(\bar{\delta}(x \otimes 1)(1 \otimes a \otimes b))$$

$$\begin{aligned} &= ((\iota \otimes \sigma)(w^* \otimes 1))(y \otimes 1)\eta_\Psi(x \otimes a \otimes b) \qquad \text{by (2.6)}, \\ &= \psi_G(d^*a)((\iota \otimes \sigma)(w^* \otimes 1))\eta_\Psi((\delta(z) \otimes 1)(x \otimes 1 \otimes b)) \quad \text{by (3.5)}, \\ &= \psi_G(d^*a)((\iota \otimes \sigma)(w^* \otimes 1)) \lim_\psi \int \eta_\Psi(((\delta_{\rho(r)*\psi}(z)x) \otimes 1)(1 \otimes \rho(r) \otimes b))dr \\ &= \psi_G(d^*a) \lim_\psi \int \eta_\Psi(\bar{\delta}((\delta_{\rho(r)*\psi}(z)x) \otimes 1)(1 \otimes \rho(r) \otimes b)) \, dr \end{aligned}$$

$$= \psi_G(d^*a) \lim_{\psi} \int \eta_\Psi(F(r,\psi))dr \qquad \text{by (3.6)},$$

$$= \psi_G(d^*a)\eta_\Psi((\iota \otimes \delta_G)(\delta(z))\bar{\delta}(x \otimes 1)(1 \otimes 1 \otimes b)) \qquad \text{by (3.7)},$$

$$= \psi_G(d^*a)(1 \otimes W_G^*)\eta_\Psi((\delta(z) \otimes 1)\bar{\delta}(x \otimes 1)(1 \otimes 1 \otimes b)) \text{ by (2.7)},$$

$$= (1 \otimes W_G^*)(y \otimes 1)\eta_\Psi(\bar{\delta}(x \otimes 1)(1 \otimes a \otimes b)) , \qquad \text{by (3.5)},$$

which gives (3.4).

(ii) ⇒ (iii): Let $\mathcal{J}$ be the set of all $y \in \bar{\mathfrak{n}}$ with (3.4). Since $\tilde{\delta} = \mathrm{Ad}_{1 \otimes W_G} \circ \bar{\delta}$, we have a co-action ζ of G on $\bar{\mathfrak{n}} \otimes F_2$ defined by

$$(3.9) \qquad \begin{cases} \zeta = \mathrm{Ad}_v \circ (\iota \otimes \iota \otimes \sigma) \circ (\bar{\delta} \otimes \iota) \\ v = (\iota \otimes \iota \otimes \sigma)(1 \otimes 1 \otimes 1 \otimes e_{11} + 1 \otimes W_G \otimes e_{22}) . \end{cases}$$

Therefore $\sum x_{ij} \otimes e_{ij} \in (\bar{\mathfrak{n}} \otimes F_2)^\zeta$ if and only if

$$x_{11} \in \bar{\mathfrak{n}}^{\bar{\delta}} , \; x_{12}^* \in \mathcal{J} , \; x_{21} \in \mathcal{J} \text{ and } x_{22} \in \bar{\mathfrak{n}}^{\tilde{\delta}} .$$

Condition (ii) implies that the central support of $1 \otimes 1 \otimes e_{11}$ in $(\bar{\mathfrak{n}} \otimes F_2)^\zeta$ is majorized by the central support of $1 \otimes 1 \otimes e_{22}$ in $(\bar{\mathfrak{n}} \otimes F_2)^\zeta$. Since $\mathfrak{n}^\delta$ is properly infinite, $1 \otimes 1 \otimes e_{22}$ is also properly infinite in $(\bar{\mathfrak{n}} \otimes F_2)^\zeta$. Since $1 \otimes 1 \otimes e_{11}$ is σ-finite in $(\bar{\mathfrak{n}} \otimes F_2)^\zeta$, $1 \otimes 1 \otimes e_{11} \prec 1 \otimes 1 \otimes e_{22}$ in $(\bar{\mathfrak{n}} \otimes F_2)^\zeta$. Thus there exists an isometry u in $\mathcal{J}$. Put $e = uu^*$. Then $e \in \bar{\mathfrak{n}}^{\bar{\delta}}$ and (iii) is established.

(iii) ⇒ (i): This follows from the similar arguments to those in the proof (iii) ⇒ (i) in Theorem 3.1. Q.E.D.

According to Theorems 3.1 and 3.2, we have a stronger result than (1.32) and (1.34).

<u>Corollary</u> 3.3 (a) If α is integrable, $\dot{\mathcal{E}}_\alpha(\mathfrak{p}_\alpha) = \mathfrak{m}^\alpha$.
(b) If δ is integrable, $\dot{\mathcal{E}}_\delta(\mathfrak{p}_\delta) = \mathfrak{n}^\delta$.

<u>Proof</u>. (a) The case where $\mathfrak{m}^\alpha$ is properly infinite: Since α is integrable, $\{\bar{\mathfrak{m}},\bar{\alpha}\} \cong \{\bar{\mathfrak{m}},\tilde{\alpha}\}_e$ for some projection $e \in \bar{\mathfrak{m}}^{\tilde{\alpha}}$. Let p be a minimal projection in $\mathcal{L}(L^2(G))$. Then $1 \otimes p \in \bar{\mathfrak{m}}^{\bar{\alpha}}$, which corresponds to a projection $f \in \bar{\mathfrak{m}}^{\tilde{\alpha}}$ through the above equivalence. Therefore

$$\{\mathfrak{m},\alpha\} = \{\bar{\mathfrak{m}},\bar{\alpha}\}_{1 \otimes p} \cong \{\bar{\mathfrak{m}},\tilde{\alpha}\}_f .$$

Since $\tilde{\alpha}$ is dual, $\overline{\mathfrak{M}}^{\tilde{\alpha}} = \dot{\mathcal{E}}_{\tilde{\alpha}}(p_{\tilde{\alpha}})$ by (1.34). Since $f \in \overline{\mathfrak{M}}^{\tilde{\alpha}}$, the induction $\overline{\mathfrak{M}}_f$ of $\overline{\mathfrak{M}}$ preserves this equality. Thus (a) is obtained.

The general case: Put $\mathfrak{P} = \mathfrak{M} \bar{\otimes} F_\infty$ and $\beta = (\iota \otimes \sigma) \circ (\alpha \otimes \iota)$ on $\mathfrak{P}$. Then β is an integrable action of G on $\mathfrak{P}$ and $\mathfrak{P}^\beta$ is properly infinite. Therefore $\dot{\mathcal{E}}_\beta(p_\beta) = \mathfrak{P}^\beta$. Take a minimal projection $p \in F_\infty$ and consider the induction $\mathfrak{P}_{1\otimes p}$ of $\mathfrak{P}$. Then (a) is proved.

(b) Similarly as above. Q.E.D.

NOTES

In Theorem 3.1, the equivalence (i) $\Longleftrightarrow$ (iii) is due to [14] and the rest is due to [47]. In Theorem 2.2. the implifications (ii) $\Rightarrow$ (iii) $\Rightarrow$ (i) are proved in [47] and the implication (i) $\Rightarrow$ (ii) is new.

§ 4. Dominant actions and co-actions.

In this section we shall introduce two new concepts "semi-dual" and "dominant" for actions and co-actions, and show the implications:

$$\text{"Dominant"} \Rightarrow \text{"Dual"} \Rightarrow \text{"Semi-Dual"} \Rightarrow \text{"Intregrable"} .$$

Definition 4.1. An action α (resp. co-action δ) is said to be semi-dual, if there exists a unitary $v \in \mathfrak{M} \bar{\otimes} \mathcal{R}(G)'$ (resp. $u \in \mathfrak{N} \bar{\otimes} L^\infty(G)$) such that

$$\bar{\alpha}(v) = (v \otimes 1)(1 \otimes V_G') \tag{4.1}$$

$$(\text{resp. } \bar{\delta}(u) = (u \otimes 1)(1 \otimes W_G)) . \tag{4.2}$$

It should be noted that if α (resp. δ) is dual, then it is semi-dual. Moreover, if α or δ is semi-dual, then it is integrable by Theorems 3.1 and 3.2.

Condition (4.2) is equivalent to the condition that for each $t \in G$ there exists a unitary $u(t) \in \mathfrak{N}$ such that $\delta(u(t)) = u(t) \otimes \rho(t)$.

From the above definition the following corollary is immediate by considering the fixed point subalgebras.

Corollary 4.2. If α and δ are semi-dual, then

$$\mathfrak{M} \times_\alpha G \cong \mathfrak{M}^\alpha \bar{\otimes} \mathcal{L}(L^2(G)) ; \tag{4.3}$$

$$\mathfrak{N} \times_\delta G \cong \mathfrak{N}^\delta \bar{\otimes} \mathcal{L}(L^2(G)) . \tag{4.4}$$

Definition 4.3. An action α (resp. co-action δ) is said to be dominant if

(i) α (resp. δ) is dual;
(ii) $\mathfrak{M}^\alpha$ (resp. $\mathfrak{N}^\delta$) is properly infinite .

In the above definition, condition (ii) implies

$$\{\mathfrak{M},\alpha\} \cong \{\bar{\mathfrak{M}},\bar{\alpha}\} \ (\text{resp. } \{\mathfrak{N},\delta\} \cong \{\bar{\mathfrak{N}},\bar{\delta}\}) . \tag{4.5}$$

If we combine this with condition (i), then

$$\{\mathfrak{M},\alpha\} \cong \{\bar{\mathfrak{M}},\tilde{\alpha}\} \ (\text{resp. } \{\mathfrak{N},\delta\} \cong \{\bar{\mathfrak{N}},\tilde{\delta}\}) \tag{4.6}$$

for any dominant α (resp. δ).

Now, we shall give an analogous, but a stronger result as Theorem II.2.1 and II.2.2.

Theorem 4.4 If $\mathfrak{M}^\alpha$ (resp. $\mathfrak{N}^\delta$) is properly infinite then the following four conditions are equivalent for α (resp. δ):

(i) It is semi-dual;

(ii) It is dual;

(iii) It is dominant;

(iv) There exists a unitary $w^K \in \mathfrak{M} \bar{\otimes} \mathcal{R}(G)'$ (resp. $u \in \mathfrak{N} \bar{\otimes} L^\infty(G)$) such that $\bar{\alpha}(w^K) = (w^K \otimes 1)(\iota \otimes V_G')$, i.e. $(\alpha_t \otimes \iota)(w^K) = w^K(1 \otimes \lambda(t))$,

$$(\text{resp. } \bar{\delta}(u) = (u \otimes 1)(\iota \otimes W_G)) .$$

Proof. (i) ⇒ (ii): Combining (4.5) and (4.1) (resp. (4.2)), we have (4.6). Thus α and δ are dual.

(ii) ⟺ (iii) and (iv) ⇒ (i): obvious.

(ii) ⇒ (iv): There exists a unitary $w \in \mathfrak{M} \bar{\otimes} \mathcal{R}(G)$ (resp. $u \in \mathfrak{N} \bar{\otimes} L^\infty(G)$) such that

$$(\alpha_t \otimes \iota)(w^*) = w^*(1 \otimes \rho(t)) \quad (\text{resp. } \bar{\delta}(u) = (u \otimes 1)(1 \otimes W_G))$$

by (II.2.7) (resp. (II.2.3)). Here, we put $(K\xi)(t) = \Delta(t)^{1/2}\xi(t^{-1})$ for $\xi \in L^2(G)$ and $w^K = (1 \otimes K)w^*(1 \otimes K)$. Then

$$(\alpha_t \otimes \iota)(w^K) = w^K(1 \otimes \lambda(t)) ,$$

which implies $w^K \in \mathfrak{M} \bar{\otimes} \mathcal{R}(G)'$ and $\bar{\alpha}(w^K) = (w^K \otimes 1)(1 \otimes V_G')$. Q.E.D.

Example. We shall give an example of an action which is semi-dual but not dual. Let G be a locally compact abelian group and $\hat{G}$ its dual group. Put $\mathfrak{M} = \mathcal{L}(L^2(G))$ and

$$(u(s)\xi)(t) = \xi(t - s) \qquad s,t \in G$$

$$(v(p)\xi)(t) = \langle t,p \rangle \xi(t) \qquad p \in G$$

for $\xi \in L^2(G)$. Then the commutation relation $u(s)v(p) = \overline{\langle s,p \rangle} v(p)u(s)$ holds. Let α be the action of $G \times \hat{G}$ on $\mathfrak{M}$ defined by

$$\alpha_{(s,p)}(x) = \mathrm{Ad}_{u(s)v(p)}(x) .$$

Then $\alpha_{(s,p)}(u(t)v(q)) = \langle (s,p),(q,t) \rangle u(t)v(q)$ and hence α is semi-dual. On the other hand, $\mathfrak{M}^\alpha = \mathbf{C}$. Therefore, if α is dual, $\mathfrak{M}$ must be isomorphic to $L^\infty(\hat{G} \times G)$ by Theorem II.2.2. This is impossible for a non trivial G.

The situation will be clearer by introducing the concept of regular extensions.

Definition 4.5. Let $\alpha_{(\cdot)} : t \in G \to \alpha_t \in \mathrm{Aut}(\mathfrak{M})$ be a Borel mapping with $t \to \alpha_t$ a homomorphism up to inner automorphisms. For each Borel family $(s,t) \in G \times G \to u(s,t) \in \mathfrak{M}$ of unitaries satisfying

$$\alpha_s \circ \alpha_t = \mathrm{Ad}_{u(s,t)} \circ \alpha_{st} ,$$

the von Neumann algebra $\mathfrak{M} \times_{\alpha,u} G$ generated by $\alpha(\mathfrak{M})$ and $\rho^u(G)$, where

$$(\alpha(x)\xi)(t) = \alpha_t(x)\xi(t) \ , \quad (\rho^u(r)\xi)(t) = u(t,r)\xi(tr) \ ,$$

is called a <u>regular</u> <u>extension</u> of $\mathfrak{M}$ by G with respect to α and u.

If we define a map $\hat{\alpha}$ on $\mathfrak{M} \times_{\alpha,u} G$ by

$$\hat{\alpha}(y) = \mathrm{Ad}_{1 \otimes W_G^*}(y \otimes 1) \ ,$$

then $\hat{\alpha}$ is a co-action of G on $\mathfrak{M} \times_{\alpha,u} G$ by direct computation. The relation between a semi-dual co-action and a regular extension is given by the following, which is also an extension of Theorem II.2.1.

<u>Theorem</u> 4.6. Let δ be a co-action of G on $\mathfrak{N}$. The following three conditions are equivalent:

(i) $\{\mathfrak{N},\delta\} \cong \{\mathfrak{M} \times_{\alpha,u} G, \hat{\alpha}\}$ for some $\mathfrak{M}$, α and u.

(ii) There exists a Borel map $t \in G \to u(t) \in \mathfrak{N}$ with unitary values such that $\delta(u(t)) = u(t) \otimes \rho(t)$, $t \in G$.

(iii) There exists a unitary u in $\mathfrak{N} \bar{\otimes} L^\infty(G)$ such that $\bar{\delta}(u) = (u \otimes 1)(1 \otimes W_G)$.

We leave the proof to the reader.

NOTES

The general theory of dominant actions is developed in [14]. The part of this section dealing with an action should be viewed as a supplement to [14; Chapter III]. A characterization of semi-dual actions is adapted from [50].

CHAPTER IV.

SPECTRUM.

Introduction. The Arveson-Connes theory of the spectrum of actions of an abelian group was very successful and played a vital role in the structure analysis of a factor of type III. In this chapter, we shall try to generalize this analysis to the noncommutative case including the dualization.

Since a co-action δ is in essence an action of a commutative object, $A(G)$ for instance, it is natural to expect that the dualization of the Arveson-Connes theory would be smoother than the non-commutative generalization. It will be seen in §1 that this is indeed the case. Namely, the spectrum $sp(\delta)$ of δ and the Connes spectrum $\Gamma(\delta)$ are introduced there. We then prove that $\Gamma(\delta)$ is a closed subgroup of G. For a dual δ, $\Gamma(\delta)$ is shown to be the kernel of the restriction of $\hat{\delta}$ on $\mathcal{C}_{\mathfrak{n}\times_\delta G}$, Theorem 1.5. An example shows that $\Gamma(\delta)$ can fail to be a normal subgroup of G. This misbehavior of $\Gamma(\delta)$ will be corrected in the next chapter by amending $\Gamma(\delta)$ to the normalized Connes spectrum $\Gamma_n(\delta)$, see Definition V. 2.5 and Theorem V.2.6.

Section 2 is devoted to a non-commutative version of the Arveson-Connes theory in the campact case. Here we employ Roberts' apparatus. The eigensubspace in the abelian case is then replaced by $\mathfrak{m}^\alpha(V)$ for a unitary representation $\{V, \mathfrak{H}_V\}$ of G, where $\mathfrak{m}^\alpha(V)$ is defined by (2.8). Making use of $\mathfrak{m}^\alpha(V)$, the monoidal spectrum $Msp(\alpha)$ is defined and serves as a non-commutative version of the spectrum of α for an abelin group.

Section 3, the connection between $\Gamma(\delta)$ and the center of $\mathfrak{n}\times_\delta G$ is given. This relation will become sharper in Theorem V.2.6. The analysis of the center of $\mathfrak{n}\times_\delta G$ is not satisfactory yet. We do need further effort in this area.

Section 4 is devoted to the analysis of the relation between co-actions and Roberts actions. It will be seen there that these two things are indeed equivalent, Theorem 4.8.

§1. The Connes spectrum of co-actions.

Following Connes' idea, [12], we shall introduce the essential spectrum of a co-action named the Connes spectrum, and show some properties. First of all, we recall the duality theorem for locally compact groups:

If $x \in \mathcal{R}(G)$ is non zero, then the following three conditions are equivalent:

(i) $x = \rho(t)$ for some $t \in G$.

(ii) $\delta_G(x) = x \otimes x$, i.e., $\langle x, \varphi\psi\rangle = \langle x,\varphi\rangle\langle x,\psi\rangle$ for all $\varphi, \psi \in A(G)$.

(iii) $W_G^*(x \otimes 1)W_G = x \otimes x$.

It is known that the annihilator of $\rho(t)$ in $A(G)$ is the regular maximal ideal of $A(G)$ and vice versa. The support $\text{supp}(x)$ of $x \in \mathcal{R}(G)$ is defined as the hull of a closed ideal $\{\varphi \in A(G) : \delta_{G,\varphi}(x) = 0\}$. Here $\delta_{G,\varphi}$ is defined by

$$\langle \delta_{G,\varphi}(x), \omega\rangle = \langle \delta_G(x), \omega \otimes \varphi\rangle , \qquad \omega \in A(G) .$$

Of course, $\text{supp}(x) = \emptyset$ if and only if $x = 0$. The only if part is a Tauberian theorem.

Let Φ_ω, $\omega \in \mathcal{M}_*$ be a linear map of $\mathcal{M}$ into $\mathcal{R}(G)$ defined by

$$\langle \Phi_\omega(y), \varphi\rangle = \langle \delta(y), \omega \otimes \varphi\rangle , \qquad \varphi \in A(G) .$$

The closure of the union of all $\text{supp}(\Phi_\omega(y))$, $\omega \in \mathcal{M}_*$ is considered as the spectrum of $y \in \mathcal{M}$ with respect to δ. In other words:

Definition 1.1. The spectrum $sp_\delta(y)$ of $y \in \mathcal{M}$ with respect to δ is the hull of the closed ideal $m_y = \{\varphi \in A(G) : \delta_\varphi(y) = 0\}$. The spectrum $sp(\delta)$ of δ is the hull of $\{\varphi \in A(G) : \delta_\varphi = 0\}$,

It is clear that $sp(\delta)$ is the closure of the union of all $sp_\delta(y)$, $y \in \mathcal{M}$. Now, we shall prepare some elementary properties of spectrums. For each closed subset E of G we denote

$$\mathcal{M}^\delta(E) = \{y \in \mathcal{M} : sp_\delta(y) \subset E\} , \tag{1.1}$$

which is a σ-weakly closed linear subspace by Lemma 1.2.ii below.

Lemma 1.2. (i) $sp_\delta(\delta_\varphi(y)) \subset sp_\delta(y) \cap \text{supp}(\varphi)$, where $\text{supp}(\varphi)$ is the closure of $\{t \in G : \varphi(t) \neq 0\}$.

(ii) $sp_\delta(y) \subset E$ if and only if $\delta_\varphi(y) = 0$ for $\varphi \in A(G)$ vanishing on some neighborhood of E.

(iii) $sp_\delta(y^*) = sp_\delta(y)^{-1}$.

(iv) $t \in sp(\delta)$ if and only if $\mathfrak{h}^{\delta}(U) \neq \{0\}$ for all compact neighborhoods U of t.

(v) If E or F is compact, then $\mathfrak{h}^{\delta}(E)\mathfrak{h}^{\delta}(F) \subset \mathfrak{h}^{\delta}(EF)$.

(vi) If E is a closed subset of G, then the projections p_E^{δ} and q_E^{δ} given by

$$p_E^{\delta} = \vee\{supp(x^*x) : x \in \mathfrak{h}^{\delta}(E)\} ,$$
$$q_E^{\delta} = \vee\{supp(xx^*) : x \in \mathfrak{h}^{\delta}(E)\}$$

are central projections of $\mathfrak{h}^{\delta}$.

<u>Proof</u>. (i) Put $z = \delta_{\varphi}(y)$. If $\delta_{\psi}(y) = 0$, then $\delta_{\psi}(z) = \delta_{\varphi}(\delta_{\psi}(y)) = 0$. Therefore $\mathfrak{m}_y \subset \mathfrak{m}_z$ and hence $sp_{\delta}(z) \subset sp_{\delta}(y)$.

If $t \notin supp(\varphi)$, there exists a $\psi \in A(G)$ such that $\psi(t) \neq 0$ and $\varphi\psi = 0$. Since $\varphi\psi = 0$ implies $\delta_{\psi}(z) = 0$, it follows that $t \notin sp_{\delta}(z)$. Thus (i) is proved.

(ii) Suppose $y \in \mathfrak{h}^{\delta}(E)$. For any neighborhood U of e and $\varphi \in A(G)$, if φ vanishes on EU, then E is disjoint from $supp(\varphi)$. Since $sp_{\delta}(\delta_{\varphi}(y))$ is empty by (i), we have $\delta_{\varphi}(y) = 0$.

Conversely, if $t \notin E$, there exists a neighborhood U of e and $\varphi \in A(G)$ such that φ vanishes on EU and $\varphi(t) \neq 0$. From our assumption it follows that $\delta_{\varphi}(y) = 0$. Since $\varphi(t) \neq 0$, we have $t \notin sp_{\delta}(y)$. Therefore $sp_{\delta}(y) \subset E$.

(iii) Let $\mathfrak{m}_t$ denote the ideal $\{\rho(t)\}^{\perp}$ of $A(G)$. If $t \in sp_{\delta}(y^*)$, then $\mathfrak{m}_{y^*} \subset \mathfrak{m}_t$. Since $(\mathfrak{m}_y)^* = \mathfrak{m}_{y^*}$ and $(\mathfrak{m}_t)^* = \mathfrak{m}_{t^{-1}}$ we have $\mathfrak{m}_y \subset \mathfrak{m}_{t^{-1}}$ and hence $t^{-1} \in sp_{\delta}(y)$. Thus $sp_{\delta}(y^*) \subset sp_{\delta}(y)^{-1}$. Changing the role of y and y^*, we have $sp_{\delta}(y) \subset sp_{\delta}(y^*)^{-1}$.

(iv) Suppose that $\mathfrak{h}^{\delta}(U) \neq \{0\}$ for all compact neighborhoods U of t. Then U has a non empty intersection with some $sp_{\delta}(y)$ with $y \in \mathfrak{h}$. Therefore t belongs to the closure of union of $sp_{\delta}(y)$ with $y \in \mathfrak{h}$, and hence to $sp(\delta)$.

Conversely, suppose that $t \in sp(\delta)$. For any compact neighborhood U of t there exists a $\varphi \in A(G)$ such that $\varphi(t) \neq 0$ and $supp(\varphi) \subset U$. Since $t \in sp(\delta)$, $\delta_{\varphi}(y) \neq 0$ for some $y \in \mathfrak{h}$. Since $sp_{\delta}(\delta_{\varphi}(y)) \subset U$ by (i), it follows that $\delta_{\varphi}(y) \in \mathfrak{h}^{\delta}(U)$.

(v) We may assume by (iii) that F is compact. Suppose that $x \in \mathfrak{h}^{\delta}(E)$ and $y \in \mathfrak{h}^{\delta}(F)$. For any $\omega \in \mathfrak{h}_*$ and $\varphi \in A(G)$ we have

$$\begin{aligned}\langle \Phi_{\omega}(xy),\varphi\rangle &= \langle \delta(x)\delta(y),\ \omega \otimes \varphi\rangle \\ &= \lim \int \langle \delta(x),\ \delta_{\rho(r)^*\psi}(y)\,\omega \otimes \rho(r)\varphi\rangle \ dr \\ &= \lim \int \langle \Phi_{\omega(r,\psi)}(x)\rho(r),\varphi\rangle \ dr ,\end{aligned}$$

where $\omega(r,\psi) = \delta_{\rho(r)^*\psi}(y)\omega$. Therefore,

$$\Phi_{\omega}(xy) = \lim \int \Phi_{\omega(r,\psi)}(x)\rho(r)\ dr .$$

Since $sp_{\delta_G}(z\rho(r)) = sp_{\delta_G}(z)r$ and $sp_{\delta_G}(\Phi_{\omega(r,\psi)}(x)) \subset sp_\delta(x)$, it follows that

$$sp_{\delta_G}(\Phi_{\omega(r,\psi)}(x)\rho(r)) \subset sp_\delta(x)r .$$

If U is a symmetric compact neighborhood of e, then the map $: r \to \delta_{\rho(r)*\psi}(y)$ vanishes on the complement of UF for all ψ with $\mathrm{supp}(\psi) \subset U$. Therefore

$$sp_{\delta_G}(\Phi_\omega(xy)) \subset sp_\delta(x)UF .$$

Since U is arbitrary, the left hand side is contained in EF. Thus $xy \in \mathfrak{n}^\delta(EF)$.

(vi) We may assume that there exists a unitary $W \in \mathcal{L}(\mathfrak{H}) \bar{\otimes} \mathfrak{R}(G)$ which implements the co-action δ. As in Appendix, there exists a $*$-representation π of $C_\infty(G)$ on $\mathfrak{H}$ such that $(\pi(\varphi)\xi|\eta) = \langle W, \omega_{\xi,\eta} \otimes \varphi\rangle$, $\xi, \eta \in \mathfrak{H}$. From the proof of Theorem A.1, it follows that W belongs to $\pi(C_\infty(G))'' \bar{\otimes} \mathfrak{R}(G)$. We claim that p_E^δ (resp. q_E^δ) belongs to $\pi(C_\infty(G))'$. If $\xi \in (1 - p_E^\delta)\mathfrak{H}$, then for any $a \in \mathfrak{n}^\delta(E)$ and $\varphi \in A(G)$ we have

$$(a\pi(\varphi)\xi|\eta) = (\pi(\varphi)\xi|a^*\eta) = \langle W, \omega_{\xi,a^*\eta} \otimes \varphi\rangle = \langle W, \omega_{\xi,\eta}a \otimes \varphi\rangle$$

$$= \langle (a \otimes 1)W, \omega_{\xi,\eta} \otimes \varphi\rangle = \langle W\delta(a), \omega_{\xi,\eta} \otimes \varphi\rangle .$$

If $\varphi = \omega_{f,g}$, then

$$(a\pi(\varphi)\xi|\eta) = (W\delta(a)(\xi \otimes f)|\eta \otimes g) .$$

On the other hand, for any $\zeta \in \mathfrak{H}$ and $h \in L^2(G)$, we have

$$(\delta(a)(\xi \otimes f)|\zeta \otimes h) = \langle \delta(a), \omega_{\xi,\zeta} \otimes \omega_{f,h}\rangle$$

$$= (\delta_\psi(a)\xi|\zeta) = 0 .$$

where $\psi = \omega_{f,h}$. This means that $\delta(a)(\xi \otimes f) = 0$. Hence we get

$$(a\pi(\varphi)\xi|\eta) = 0 , \qquad \eta \in \mathfrak{H} .$$

Thus, $a\pi(\varphi)\xi = 0$ for every $a \in \mathfrak{n}^\delta(E)$, so that $\pi(\varphi)\xi$ belongs to $(1 - p_E^\delta)\mathfrak{H}$. Hence $\pi(\varphi)$ leaves $(1 - p_E^\delta)\mathfrak{H}$ is variant, which means that $p_E^\delta \in \pi(A(G))'$.

Since $W \in \pi(A(G))''\bar{\otimes}\mathfrak{R}(G)$, we have

$$\delta(p_E^\delta) = W^*(p_E^\delta \otimes 1)W = p_E^\delta \otimes 1 .$$

The assertion for q_E^δ follows from the fact that $p_{E^{-1}}^\delta = q_E^\delta$.

Next, $\mathfrak{n}^\delta(E)$ is a two-sided module over $\mathfrak{n}^\delta$. Hence p_E^δ and q_E^δ are both invariant under the inner automorphisms induced by unitaries of $\mathfrak{n}^\delta$. Hence p_E^δ and q_E^δ are both in the center of $\mathfrak{n}^\delta$. Q.E.D.

For each projection e in $\mathcal{N}^\delta$ we define a co-action δ^e of G on $\mathcal{N}_e$ as the restriction of δ to $\mathcal{N}_e$. It then follows that if f is the central support e in $\mathcal{N}^\delta$, then $sp(\delta^e) = sp(\delta^f)$.

Definition 1.3. The Connes spectrum $\Gamma(\delta)$ of δ is the intersection of all $sp(\delta^e)$, where e runs over all non-zero projections in $\mathcal{N}^\delta$.

Theorem 1.4. The Connes spectrum $\Gamma(\delta)$ is a closed subgroup of G.

Proof. Since $\Gamma(\delta)$ is clearly closed, it suffices to show the group property. Since $\Gamma(\delta)^{-1} = \Gamma(\delta)$ by (iii) of Lemma 1.2, we have only to show $\Gamma(\delta)sp(\delta^e) \subset sp(\delta^e)$ for all projections $e \in \mathcal{N}^\delta$. For a $t \in G$ to be in $\Gamma(\delta)$ it is necessary and sufficient that for any compact neighborhood U of t and a non-zero projection $e \in \mathcal{N}^\delta$, $p_U^{\delta^e} = e$ and $q_U^{\delta^e} = e$. Hence for any $s \in sp(\delta^e)$ and a compact neighborhood V of s we have

$$\mathcal{N}^{\delta^e}(UV) \supset \mathcal{N}^{\delta^e}(U)\mathcal{N}^{\delta^e}(V) \neq \{0\} .$$

Hence $UV \cap sp(\delta^e) \neq \emptyset$. Since $\{UV\}$ forms a basis for neighborhoods of ts, we have $ts \in sp(\delta^e)$. Q.E.D.

Theorem 1.5. If δ is dual, then
(i) $\Gamma(\delta) = \{t \in G : \hat{\delta}_t = \iota$ on $\mathcal{C}_{\mathcal{N} \times_\delta G}\}$, and
(ii) $\Gamma(\delta) = \Gamma(\hat{\hat{\delta}})$.
In particular, $\Gamma(\hat{\alpha}) = \{t \in G : \alpha_t = \iota$ on $\mathcal{C}_{\mathcal{M}}\}$.

Proof. (i) Suppose that δ is dual. There exists an isomorphism π of $\mathcal{N}^\delta \times_\alpha G$ onto $\mathcal{N}$ satisfying $\delta \circ \pi = (\pi \otimes \iota) \circ \hat{\alpha}$, where α is an action of G on $\mathcal{N}^\delta$. Then the duality of crossed products for action tells us that

$$\{\mathcal{N} \times_\delta G, \hat{\delta}\} \cong \{\mathcal{N}^\delta \bar{\otimes} \mathcal{L}(L^2(G)), \tilde{\alpha}\} .$$

Therefore

$$\text{Ker}\, \hat{\delta} \restriction \mathcal{C}_{\mathcal{N} \times_\delta G} = \text{Ker}\, \tilde{\alpha} \restriction \mathcal{C}_{\mathcal{N}^\delta \otimes \mathbf{C}} = \text{Ker}\, \alpha \restriction \mathcal{C}_{\mathcal{N}^\delta} . \tag{1.2}$$

We then define an action β of G on $\mathcal{N}^\delta$ by:

$$\beta_t \circ (\pi \circ \alpha) = (\pi \circ \alpha) \circ \alpha_t .$$

Here we set $u(t) = \pi(1 \otimes \rho(t))$. Then $\beta_t = \text{Ad}_{u(t)}$. Since $\delta(u(t)) = u(t) \otimes \rho(t)$, $\mathcal{N}^\delta(\{t\}) = \mathcal{N}^\delta u(t)$. If e is a non zero central projection in $\mathcal{N}^\delta$, then $\beta_t(e) = u(t)eu(t)^*$ and

$$e\mathcal{N}^\delta(\{t\})e = e\mathcal{N}^\delta u(t)e = e\mathcal{N}^\delta \beta_t(e)u(t) . \tag{1.3}$$

Consequently, $\hat{\delta}_s = \iota$ on the center of $\mathcal{N} \times_\delta G$ if and only if $\alpha_s = \iota$ on $\mathcal{C}_{\mathcal{N}^\delta}$, if and only if $\beta_s = \iota$ on $\mathcal{C}_{\mathcal{N}^\delta}$, if and only if $e\beta_s(e) \neq 0$ for all non zero projection $e \in \mathcal{C}_{\mathcal{N}^\delta}$ if and only if $s \in \Gamma(\delta)$ by (1.3).

(ii) Since δ is dual, $\{\bar{\mathcal{N}}, \bar{\delta}\} = \{\bar{\mathcal{N}}, \tilde{\delta}\}$ by Theorem II.2.1. Therefore $\Gamma(\bar{\delta}) =$

$\Gamma(\tilde{\delta}) = \Gamma(\hat{\hat{\delta}})$ by the duality for crossed product. It remains to show that $\Gamma(\delta) = \Gamma(\bar{\delta})$.

For any $y \in \mathfrak{n}$, $z \in \mathfrak{L}(L^2(G))$, $\omega_1 \in \mathfrak{n}_*$, $\omega_2 \in \mathfrak{L}(L^2(G))_*$ and $\varphi \in A(G)$, we have

$$\langle (\bar{\delta})_\varphi (y \otimes z), \omega_1 \otimes \omega_2 \rangle = \langle \delta_\varphi(y) \otimes z , \omega_1 \otimes \omega_2 \rangle .$$

Thus $\delta_\varphi = 0$ is equivalent to $(\bar{\delta})_\varphi = 0$. Since $\mathcal{C}_{\bar{\mathfrak{n}}^{\bar{\delta}}} = \mathcal{C}_{\mathfrak{n}^\delta} \otimes C$, it follows that $sp(\delta^e) = sp(\bar{\delta}^{e\otimes 1})$ for $e \in \mathcal{C}_{\mathfrak{n}^\delta}$. Thus $\Gamma(\delta) = \Gamma(\bar{\delta})$. Q.E.D.

It should be noted that, if δ is dual, then $\Gamma(\delta)$ is a normal subgroup of G. For general δ, $\Gamma(\delta)$ is not necessarily normal and so $\Gamma(\delta) \neq \Gamma(\hat{\hat{\delta}})$, unlike the abelian case, as seen in the following:

Example. If H is a non-normal closed subgroup of G, then the restriction δ of δ_G to $\mathfrak{n} = \rho(H)''$ is an integrable co-action of G. Since $\mathfrak{n}^\delta = C$, $\Gamma(\delta) = sp(\delta) = H$. On the other hand, $\Gamma(\hat{\hat{\delta}}) = \ker \hat{\delta}|_{\mathcal{C}_{\mathfrak{n}\times_\delta G}} = \bigcap_{t\in G} tHt^{-1}$. Indeed, $\mathfrak{n} \times_\delta G$ is generated by $\rho(t) \otimes \rho(t)$, $t \in H$, and $C \otimes L^\infty(G)$. Since $Ad_{V_G}(\rho(t) \otimes \rho(t)) = 1 \otimes \rho(t)$ and $Ad_{V_G}(1 \otimes f) = 1 \otimes f$, we have $\mathcal{C}_{\mathfrak{n}\times_\delta G} = C \otimes \mathfrak{L}^\infty(G/H)$. Since $\hat{\delta}_r = \iota \otimes \lambda_r$ on $\mathfrak{n} \times_\delta G$, $r \in \Gamma(\hat{\hat{\delta}})$ is equivalent to $r^{-1}tH = tH$ for every $t \in G$.

Corollary 1.6. The following three (resp. two) conditions are equivalent:

(i) $\mathcal{C}_{\mathfrak{n}\times_\delta G} \subset \delta(\mathfrak{n})$ (resp. $\mathcal{C}_{\mathfrak{m}\times_\alpha G} \subset \alpha(\mathfrak{m})$).

(ii) $\mathcal{C}_{\bar{\mathfrak{n}}\times_{\tilde{\delta}} G} \subset \tilde{\delta}(\bar{\mathfrak{n}})$ (resp. $\mathcal{C}_{\bar{\mathfrak{m}}\times_{\tilde{\alpha}} G} \subset \tilde{\alpha}(\bar{\mathfrak{m}})$).

(iii) $\Gamma(\tilde{\delta}) = G$.

Proof. The equivalence of (ii) and (iii) is immediate from Theorem 1.5.

(i) $\Leftrightarrow$ (ii): The case of δ. Put $\mathfrak{m} = \mathfrak{n} \times_\delta G$ and $\alpha = \hat{\delta}$. According to the duality for crossed product we have only to show

$$\mathcal{C}_\mathfrak{m} \subset \mathfrak{m}^\alpha \Longleftrightarrow \mathcal{C}_{\bar{\mathfrak{m}}} \subset \bar{\mathfrak{m}}^{\tilde{\alpha}} . \tag{1.4}$$

Since $\mathcal{C}_{\bar{\mathfrak{m}}} = \mathcal{C}_\mathfrak{m} \otimes C$, for any $x \in \mathcal{C}_\mathfrak{m}$ we have

$$x \in \mathfrak{m}^\alpha \Longleftrightarrow x \otimes 1 \in \bar{\mathfrak{m}}^{\tilde{\alpha}} ,$$

which implies (1.4).

The case of α. The proof can be done similarly as above by noticing that for any $y \in \mathcal{C}_\mathfrak{n}$

$$y \in \mathfrak{n}^\delta \Longleftrightarrow y \otimes 1 \in \bar{\mathfrak{n}}^{\tilde{\delta}} ,$$

because $Ad_{1\otimes W_G} \circ (\iota \otimes \sigma) \circ (\delta \otimes \iota)(y \otimes 1) = y \otimes 1 \otimes 1.$ Q.E.D.

NOTES

The spectral analysis for α in the abelian case was developed by Arveson [4], which had considerable impact on the subsequent development of the structure theory of factors of type III. The spectrum of δ was introduced by Nakagami [46]. The Connes spectrum $\Gamma(\alpha)$ was introduced by Connes [12] and used to classify factors of type III into those of type III_λ, $0 \leq \lambda \leq 1$. The relation between $\Gamma(\alpha)$ and $\mathfrak{M} \times_\alpha G$ in the abelian case was determined in [14]. Theorems 1.4 and 1.5 are due to Nakagami [47].

§2. Spectrum of actions.

If G is abelian, the spectrum of an action of G on $\mathfrak{M}$ coincides with the spectrum of a co-action $\delta = \mathrm{Ad}_{1\otimes\mathfrak{F}} \circ \alpha$ of $\hat{G}$ on $\mathfrak{M}$, where $\mathfrak{F}$ is the Fourier transform of $L^2(G)$ onto $L^2(\hat{G})$. However, if G is non abelian, the situation is unclear yet. We shall try to clarify the situation in this section.

Let $C^*(G)$ be the enveloping C^*-algebra of the involutive Banach algebra $L^1(G)$, and $W^*(G)$ the universal enveloping von Neumann algebra of $C^*(G)$, namely, the second dual of $C^*(G)$.

Now, for each $f \in L^1(G)$ we define α_f by

$$\alpha_f(x) = \int f(t)\alpha_t(x)dt .$$

The set m_α of all $f \in L^1(G)$ satisfying $\alpha_f = 0$ is a closed two sided ideal of $L^1(G)$. Since the σ-weak closure of m_α in $W^*(G)$ is also a closed two sided ideal, it is of the form $W^*(G)e_\alpha$ for some central projection e_α in $W^*(G)$.

Definition 2.1. The spectrum $sp(\alpha)$ and the essential spectrum $\Gamma(\alpha)$ are defined by

(2.1) $$sp(\alpha) = 1 - e_\alpha \quad ;$$

(2.2) $$\Gamma(\alpha) = \inf\{sp(\alpha^e) : e \in \mathfrak{M}^\alpha,\ e \neq 0\} .$$

To see the situation more closely, we shall assume throughout the rest of this section that G is compact. Let $\hat{G}$ be the equivalence classes of all irreducible unitary representations of G, χ_p, $p \in \hat{G}$ the normalized character of G and E_p, $p \in \hat{G}$ the corresponding central projection in $\mathcal{R}(G)$:

$$E_p = \int_G \chi_p(t)\rho(t)dt .$$

Then $C_{\mathcal{R}(G)} = \sum^{\oplus}_{p\in\hat{G}} \mathbf{C}E_p$.

Proposition 2.2. If $\{U,\mathfrak{H}_U\}$ is a representative of $p \in \hat{G}$, then the following four conditions are equivalent:

(i) $E_p \leq sp(\alpha)$.

(ii) $\alpha_f \neq 0$ for some $f \in E_pL^2(G)$.

(iii) $(\alpha_t \otimes \iota)(X) = X(1 \otimes U(t))$ for some non zero $X \in \mathfrak{M} \bar{\otimes} \mathcal{L}(\mathfrak{H}_U)$.

(iv) $\{U,\mathfrak{H}_U\}$ is equivalent to a subrepresentation of $\{\alpha,\mathfrak{M}\}$.

Proof. (i) $\Longleftrightarrow$ (ii): As E_p is a minimal projection in the center of $\mathcal{R}(G)$, condition (i) is equivalent to $E_p(1-e_\alpha) \neq 0$. It is clear that $\alpha_f = 0$ for all $f \in E_pL^2(G)$ if and only if $E_pL^2(G) \subset m_\alpha$, if and only if $E_p \leq e_\alpha$. Therefore, (i) is equivalent to (ii).

(ii) $\Rightarrow$ (iii): Let $d = \dim \mathfrak{H}_U$ and $\{\varepsilon_1,\dots,\varepsilon_d\}$ an orthonormal basis of $\mathfrak{H}_U$. Suppose that $\alpha_f \neq 0$ for some $f \in E_pL^2(G)$. There exists an $x \in \mathfrak{M}$ with $\alpha_f(x) \neq 0$.

Therefore

(2.3) $$x_{jk} = \int (U(s)\varepsilon_j|\varepsilon_k)^{-}\alpha_s(x)ds$$

is non zero for some j,k. Let $T_{\xi,\eta}$ denote the operator on $\mathfrak{H}_U$ defined by $T_{\xi,\eta}\zeta = (\zeta|\eta)\xi$, and

(2.4) $$X = \sum_{j,k} x_{jk} \otimes T_{\varepsilon_j,\varepsilon_k} .$$

Then $X \neq 0$, $X \in \mathfrak{M} \overline{\otimes} \mathfrak{L}(\mathfrak{H}_U)$ and

(2.5) $$(\alpha_t \otimes \iota)(X) = X(1 \otimes U(t)) .$$

by direct computation.

(iii) ⇒ (ii): Suppose that a non zero $X \in \mathfrak{M} \overline{\otimes} \mathfrak{L}(\mathfrak{H}_U)$ satisfies condition (iii). Then it is of the form

(2.6) $$X = \sum_{j,k} X_{jk} \otimes T_{\varepsilon_j,\varepsilon_k}$$

by fixing an orthonormal basis $\{\varepsilon_1,\dots,\varepsilon_d\}$ of $\mathfrak{H}_U$. Condition (iii) implies

(2.7) $$\alpha_t(X_{jk}) = \sum_{\ell} (U(t)\varepsilon_k|\varepsilon_\ell)X_{j\ell}$$

and hence

$$X_{jk} = \int (U(t)\varepsilon_k|\varepsilon_k)^{-}\alpha_t(X_{jk})dt .$$

Since $X \neq 0$, $X_{jk} \neq 0$ for some j,k, which implies (ii).

(iii) ⇒ (iv): Using the same notations as in the proof of (iii) ⇒ (ii), we fix some j with $X_{jk} \neq 0$. Put $a_k = X_{jk}$ for $k = 1,\dots,d$. Let V be a linear map of $\mathfrak{H}_U$ to the subspace of $\mathfrak{M}$ spanned by $\{a_k : k = 1,\dots,d\}$ such that $V\varepsilon_k = a_k$. Then (2.7) implies

$$\alpha_t V\varepsilon_k = \sum_{\ell} (U(t)\varepsilon_k|\varepsilon_\ell)V\varepsilon_\ell = VU(t)\varepsilon_k .$$

(iv) ⇒ (iii): Suppose that $\{U,\mathfrak{H}_U\}$ is equivalent to a subrepresentation of $\{\alpha,\mathfrak{M}\}$. Let V be the intertwiner of $\mathfrak{H}_U$ onto the subspace of $\mathfrak{M}$ on which U is carried. Let $\{\varepsilon_1,\dots,\varepsilon_d\}$ be an orthonormal basis of $\mathfrak{H}_U$ and $a_k = V\varepsilon_k$ for $k = 1,\dots,d$. Then

$$\alpha_t(a_k) = VU(t)\varepsilon_k = \sum_{\ell} (U(t)\varepsilon_k|\varepsilon_\ell)a_\ell .$$

Put $X_{jk} = a_k$ for all j,k. Then X defined by (2.6) satisfies $(\alpha_t \otimes \iota)(X) = X(1 \otimes U(t))$ and $X \neq 0$. Q.E.D.

For each irreducible unitary representation $\{U,\mathfrak{H}_U\}$ of G we denote:

(2.8) $$\mathfrak{M}^{\alpha}(U) = \{x \in \mathfrak{M} \bar{\otimes} \mathcal{L}(\mathfrak{H}_U) : (\alpha_t \otimes \iota)(x) = x(1 \otimes U(t)),\ t \in G\} ,$$

which plays the role of eigenspace associated with $\{U,\mathfrak{H}_U\}$.

Proposition 2.3. Assume that $\mathfrak{M}^{\alpha}$ is properly inifinite. If $\{U,\mathfrak{H}_U\}$ is a unitary representation of G, then the following two conditions are equivalent:

(i) There exists a representation $\{\alpha,\mathfrak{K}\}$ in $\mathcal{H}_{\alpha}(\mathfrak{M})$ equivalent to $\{U,\mathfrak{H}_U\}$.

(ii) There exists a unitary in $\mathfrak{M}^{\alpha}(U)$.

Proof. (i) $\Rightarrow$ (ii): Suppose that $\{\alpha,\mathfrak{K}\} \cong \{U,\mathfrak{H}_U\}$. Let $\mathfrak{H}_o$ be a Hilbert space in $\mathfrak{M}^{\alpha}$ with $\dim \mathfrak{H}_o = \dim \mathfrak{K}$. Let $\{v_1,\dots,v_d\}, \{w_1,\dots,w_d\}$ and $\{\varepsilon_1,\dots,\varepsilon_d\}$ be orthonormal bases of $\mathfrak{H}_o$, $\mathfrak{K}$ and $\mathfrak{H}_U$, respectively. We may assume that each w_j corresponds to ε_j in the above equivalence:

(2.9) $$(U(t)\varepsilon_j \mid \varepsilon_k) = w_k^*\alpha_t(w_j) .$$

Now, we define an isometry V of $\mathfrak{H}$ onto $\mathfrak{H} \otimes \mathfrak{H}_U$:

(2.10) $$V\xi = \sum v_j^*\xi \otimes \varepsilon_j , \quad \xi \in \mathfrak{H} .$$

Then $V^{-1}\sum \xi_j \otimes \varepsilon_j = \sum v_j\xi_j$. Here we set

$$W = V\sum w_jv_j^*\, V^{-1} .$$

Then W is a unitary and $\mathfrak{H} \otimes \mathfrak{H}_U$:

$$W = \sum v_j^*w_k \otimes T_{\varepsilon_j,\varepsilon_k} \in \mathfrak{M} \bar{\otimes} \mathcal{L}(\mathfrak{H}_U) .$$

Using (2.9), we have

$$\begin{aligned}(\alpha_t \otimes \iota)(W) &= \sum v_j^*\alpha_t(w_\ell) \otimes T_{\varepsilon_j,\varepsilon_\ell} \\ &= \sum v_j^*w_kw_k^*\alpha_t(w_\ell) \otimes T_{\varepsilon_j,\varepsilon_\ell} \\ &= \sum v_j^*w_k \otimes T_{\varepsilon_j,U(t)^*\varepsilon_k} \\ &= W(1 \otimes U(t)) .\end{aligned}$$

(ii) $\Rightarrow$ (i): Suppose that W is a unitary in $\mathfrak{M}^{\alpha}(U)$. Let $\mathfrak{H}_o$ be a Hilbert space in $\mathfrak{M}^{\alpha}$ with $\dim \mathfrak{H}_o = \dim \mathfrak{H}_U$ and let $\{v_1,\dots,v_d\}$ and $\{\varepsilon_1,\dots,\varepsilon_d\}$

are orthonormal bases of $\mathfrak{H}_0$ and $\mathfrak{H}_U$, respectively. Define an isometry V of $\mathfrak{H}$ onto $\mathfrak{H} \otimes \mathfrak{H}_U$ by (2.10). Put

$$w_j = V^{-1} W V v_j .$$

Then $\{w_1,\dots,w_d\}$ is a basis of a Hilbert space $\mathfrak{K}$ in $\mathcal{H}_\alpha(\mathfrak{M})$. Indeed, since

$$(\alpha_t \otimes \iota) \circ \mathrm{Ad}_V = \mathrm{Ad}_V \circ \alpha_t$$

$$V^{-1}(1 \otimes U(t))V = \sum_{j,k} (U(t)\varepsilon_j \mid \varepsilon_k) v_k v_j^*$$

by direct computation, it follows that

$$(2.11)\quad \begin{aligned} \alpha_t(w_j) &= \alpha_t(V^{-1}WV)v_j = (V^{-1}WV)(V^{-1}(1 \otimes U(t))V)v_j \\ &= \sum_k (U(t)\varepsilon_j \mid \varepsilon_k) V^{-1}WV_{v^k} = \sum_k (U(t)\varepsilon_j \mid \varepsilon_k) w_k . \end{aligned}$$

Moreover, (2.11) shows that $\{U,\mathfrak{H}_U\} \cong \{\alpha,\mathfrak{K}\}$. Q.E.D.

The monoidal spectrum $\mathrm{Msp}(\alpha)$ due to Roberts is the set of all $p \in \hat{G}$ such that $\mathfrak{M}^\alpha(U)$ contains a unitary associated with a representative $\{U,\mathfrak{H}_U\}$ of p. The following theorem will give a sufficient condition for $\mathrm{Msp}(\alpha) = \{p \in \hat{G} : E_p \leq \mathrm{sp}(\alpha)\}$.

Proposition 2.4. Assume that $\mathfrak{M}^\alpha$ is properly infinite. If there exists an ergodic subgroup $\mathcal{S}$ of $\mathrm{Aut}(\mathfrak{M})$ commuting with α_t, $t \in G$, then

$$\mathrm{Msp}(\alpha) = \{p \in \hat{G} : E_p \leq \mathrm{sp}(\alpha)\} .$$

Proof. By Proposition 2.2, $E_p \leq \mathrm{sp}(\alpha)$ is equivalent to $\mathfrak{M}^\alpha(U) \neq \{0\}$. Therefore, it suffices to show that $E_p \leq \mathrm{sp}(\alpha)$ implies $p \in \mathrm{Msp}(\alpha)$. For each $\tau \in \mathcal{S}$ we set $\bar{\tau} = \tau \otimes \iota$ on $\mathfrak{M} \bar{\otimes} \mathcal{L}(\mathfrak{H}_U)$. Then the left support z_1 and the right support z_2 of $\mathfrak{M}^\alpha(U)$ are both invariant under $\bar{\tau}$. Since $\mathcal{S}$ is ergodic on $\mathfrak{M}$, z_1 and z_2 belong to $\mathbf{C} \otimes \mathcal{L}(\mathfrak{H}_U)$. Put $\mathcal{P}_U = \mathfrak{M} \bar{\otimes} \mathcal{L}(\mathfrak{H}_U)$, $\bar{\alpha}_t = \alpha_t \otimes \iota$ and $\tilde{\alpha}_t = \alpha_t \otimes \mathrm{Ad}_{U(t)}$. Since

$$(\mathcal{P}_U)^{\bar{\alpha}} \mathfrak{M}^\alpha(U) (\mathcal{P}_U)^{\tilde{\alpha}} \subset \mathfrak{M}^\alpha(U) ,$$

z_1 and z_2 are central in $(\mathcal{P}_U)^{\bar{\alpha}}$ and $(\mathcal{P}_U)^{\tilde{\alpha}}$, respectively. Thus, $z_1 = 1$. Since $\mathrm{Ad}_{1 \otimes U(t)}(z_2) = \tilde{\alpha}_t(z_2) = z_2$ and U is irreducible, $z_2 = 1$. Since $\mathfrak{M}^\alpha$ is properly infinite, we can choose a unitary in $\mathfrak{M}^\alpha(U)$. Q.E.D.

Proposition 2.5. Assume that $\mathfrak{M}^\alpha$ is properly infinite. If $\mathrm{Msp}(\alpha) = \{p \in \hat{G} : E_p \leq \mathrm{sp}(\alpha)\}$ and α is faithful, then α is dominant.

Proof. As shown in Theorem I.3.4, the set $C_\alpha(G)$ of all functions $f_{x,y}$ defined by

$$f_{x,y}(t) = x^*\alpha_t(y), \quad x,y \in \mathfrak{K}, \ \mathfrak{K} \in \mathcal{H}_\alpha(\mathfrak{M}) ,$$

is a *-subalgebra of $C(G)$. Let H be the closed subgroup of $t \in G$ such that $f_{x,y}(t) = f_{x,y}(e)$ for all $x,y \in \mathfrak{K}$ and $\mathfrak{K} \in \mathcal{H}_\alpha(\mathfrak{M})$. Then $C_\alpha(G)^- = \{f \in C(G) : \lambda_t(f) = f, \ t \in H\}$. Moreover, if $t \in H$, then $\alpha_t = \iota$ on $\mathfrak{K} \in \mathcal{H}_\alpha(\mathfrak{M})$. Since $\mathrm{Msp}(\alpha) = \{p \in G : E_p \leq \mathrm{sp}(\alpha)\}$, $\alpha_t = \iota$ on $\mathfrak{M}$. Since α is faithful, $H = \{e\}$ and so $C_\alpha(G)^- = C(G)$. The Peter-Weyl theorem then implies that $\mathcal{H}_\alpha(\mathfrak{M})$ contains a representative of p for all $p \in \hat{G}$. Thus α is dominant. Q.E.D.

Proposition 2.5. If $\{\rho,\eta\}$ is a Roberts action of $\mathcal{H}_\alpha(\mathfrak{M})$ on $\mathfrak{h}$ and $\{\alpha,\mathfrak{K}\} \in \mathcal{H}_\alpha(\mathfrak{M})$ is a representative of $p \in \hat{G}$, then the following two conditions are equivalent.

(i) $E_p \leq \Gamma(\alpha)$.

(ii) $e\rho_{\mathfrak{K}}(e) \neq 0$ for all non zero projection $e \in C_{\mathfrak{M}^\alpha}$

Proof. Condition (i) is equivalent to

$$(e \otimes 1)\mathfrak{M}^\alpha(U)(e \otimes 1) \neq 0$$

for all $e \in C_{\mathfrak{M}^\alpha}$, where $\{U,\mathfrak{H}\} \cong \{\alpha,\mathfrak{K}\}$. Q.E.D.

Proposition 2.6. If α is dual and $\mathfrak{M}^\alpha$ is properly infinite, then $\mathfrak{M}$ is generated by $\mathfrak{M}^\alpha$ and $\mathcal{H}_\alpha(\mathfrak{M})$.

Lemma 2.7. If α is semi-dual, then for any irreducible unitary representation $\{\pi,\mathfrak{H}_\pi\}$ of G there exists a unitary in $\mathfrak{M}^\alpha(\pi)$.

Proof. If α is semi-dual, there exists a unitary w in $\mathfrak{M} \bar{\otimes} \mathfrak{R}(G)$ such that $(\alpha_t \otimes \iota)(w) = w(1 \otimes \rho(t))$. If $\{\pi,\mathfrak{H}_\pi\}$ is an irreducible unitary representation of G, there exists a projection e in $\mathfrak{R}(G)'$ such that

$$\{\pi,\mathfrak{H}_\pi\} = u\{\rho,L^2(G)\}_e u^*$$

for some isometry u of $eL^2(G)$ onto $\mathfrak{H}_\pi$. Since $e \in \mathfrak{R}(G)'$, $v = (1 \otimes u)w(1 \otimes u^*)$ is a unitary in $\mathfrak{M} \bar{\otimes} \mathfrak{L}(\mathfrak{H}_\pi)$ and $(\alpha_t \otimes \iota)(v) = v(1 \otimes \pi(t))$. Q.E.D.

Proof of Proposition 2.6. Combining Proposition 2.3 and Lemma 2.7, we have for any irreducible unitary representation $\{U,\mathfrak{H}_U\}$ of G an equivalent element $\{\alpha,\mathfrak{K}\}$ in $\mathcal{H}_\alpha(\mathfrak{M})$. Let $\{\varepsilon_1,\dots,\varepsilon_d\}$ and $\{v_1,\dots,v_d\}$ be the corresponding orthonormal bases in $\mathfrak{H}_U$ and $\mathfrak{K}$, respectively. If $x \in \mathfrak{M}$, then

$$\int (U(t)\varepsilon_j \mid \varepsilon_k)^- \alpha_t(x)dt = \mathcal{E}_\alpha(xv_j^*)v_k . \tag{2.12}$$

Let p be an element in $\hat{G}$ associated with $\{U, \mathfrak{H}_U\}$. The set of functions $f_{\xi,\eta}(t) = (U(t)\xi \mid \eta)$ does not depend on the choice of a representative of p, which will be denoted by $f^p_{\xi,\eta}$. The linear span of such functions with $\xi, \eta \in \mathfrak{H}_U$ and $p \in \hat{G}$, is norm dense in $\mathfrak{C}(G)$ by Stone-Weierstrass theorem. Thus x is approximated by linear combinations of elements of the form (2.12). Consequently, $\mathfrak{M}$ is generated by $\mathfrak{M}^\alpha$ and $\mathfrak{H}_\alpha(\mathfrak{M})$. Q.E.D.

NOTES

Propositions 2.6 and 2.8 are due to Roberts [55] and Propositions 2.2, 2.4 and 2.5 are taken from [3].

§3. The center of a crossed product and $\Gamma(\delta)$.

In this section we shall give a necessary and sufficient condition for a crossed product to be a factor. The following proposition gives us a sufficient condition for the fixed point algebra to be a factor.

Proposition 3.1. If α(resp. δ) is dual, then the following two conditions are equivalent:

(i) $\mathcal{C}_{\mathfrak{M}\times_\alpha G} \subset \alpha(\mathfrak{M})$, (resp. $\Gamma(\delta) = G$);

(ii) $\mathcal{C}_{\mathfrak{M}^\alpha} \subset \mathcal{C}_{\mathfrak{M}}$, (resp. $\mathcal{C}_{\mathfrak{N}^\delta} \subset \mathcal{C}_{\mathfrak{N}}$).

Proof. The case of $\{\mathfrak{M},\alpha\}$: Since α is dual, $\{\mathfrak{M},\alpha\} \cong \{\mathfrak{N} \times_\delta G, \hat{\delta}\}$ for some $\{\mathfrak{N},\delta\}$. By means of the duality theorem for crossed product and Theorem II.1.1, it suffices to show the equivalence:

$$(3.1) \qquad \mathcal{C}_{\mathfrak{N}} \otimes \mathbf{C} \subset \mathfrak{N} \times_\delta G \Longleftrightarrow \delta(\mathcal{C}_{\mathfrak{N}}) \subset \mathcal{C}_{\mathfrak{N}\times_\delta G} .$$

Since $\mathcal{C}_{\mathfrak{N}} \otimes \mathbf{C}$ is elementwise $\hat{\delta}$ invariant, the condition in the left hand side of (3.1) is equivalent to $\mathcal{C}_{\mathfrak{N}} \otimes \mathbf{C} \subset \delta(\mathfrak{N})$, which is equivalent to

$$(3.2) \qquad \delta(\mathcal{C}_{\mathfrak{N}}) \subset (\mathfrak{N} \times_\delta G)' .$$

Indeed, $\mathcal{C}_{\mathfrak{N}} \otimes \mathbf{C} \subset \delta(\mathfrak{N})$ is equivalent to $\mathcal{C}_{\mathfrak{N}} \subset \mathfrak{N}^\delta$, for

$$(\delta \otimes \iota)(z \otimes 1) = z \otimes 1 \otimes 1 , \qquad z \in \mathcal{C}_{\mathfrak{N}} .$$

Moreover, $\mathcal{C}_{\mathfrak{N}} \subset \mathfrak{N}^\delta$ is equivalent to (3.2), for (3.2) implies $[\delta(z), 1 \otimes f] = 0$ for all $z \in \mathcal{C}_{\mathfrak{N}}$ and $f \in L^\infty(G)$. Condition (3.2) is clearly equivalent to the condition on the right hand side of (3.1).

The case of $\{\mathfrak{N},\delta\}$: Since $\{\mathfrak{N},\delta\} \cong \{\mathfrak{M} \times_\alpha G, \hat{\alpha}\}$ for some $\{\mathfrak{M},\alpha\}$. By the duality theorem for crossed product and Theorem II .1.1, it suffices to show the equivalence:

$$(3.3) \qquad \mathcal{C}_{\mathfrak{M}} \otimes \mathbf{C} \subset \mathfrak{M} \times_\alpha G \Longleftrightarrow \alpha(\mathcal{C}_{\mathfrak{M}}) \subset \mathcal{C}_{\mathfrak{M}\times_\alpha G} .$$

This is done as a similar argument as in the above case. That is, the left hand side of (3.3) is equivalent to $\mathcal{C}_{\mathfrak{M}} \otimes \mathbf{C} \subset \alpha(\mathfrak{M})$, which is equivalent to $\mathcal{C}_{\mathfrak{M}} \subset \mathfrak{M}^\alpha$, which is equivalent to $\alpha(\mathcal{C}_{\mathfrak{M}}) \subset (\mathfrak{M} \times_\alpha G)'$, which is equivalent to the right hand side of (3.3).

Q.E.D.

Corollary 3.2. Assume that $\mathcal{C}_{\mathfrak{M}^\alpha} \subset \mathcal{C}_{\mathfrak{M}}$ or $\mathcal{C}_{\mathfrak{M}\times_\alpha G} \subset \alpha(\mathfrak{M})$. (resp. $\mathcal{C}_{\mathfrak{N}^\delta} \subset \mathcal{C}_{\mathfrak{N}}$ or $\Gamma(\delta) = G$). If α (resp. δ) is dual, so is α^e (resp. δ^e) for each central projection e in $\mathfrak{M}^\alpha$ (resp. $\mathfrak{N}^\delta$).

Proof. By Theorem II .2.1 (resp. II.2.2) and Proposition 3.1. Q.E.D.

Theorem 3.3. The following two conditions are equivalent:

(i) $\mathfrak{M} \times_\alpha G$ (resp. $\mathfrak{N} \times_\delta G$) is a factor.

(ii) $C_{\mathfrak{M}\times_\alpha G} \subset \alpha(\mathfrak{M})$ (resp. $\Gamma(\delta) = G$) and α(resp. δ) is ergodic on the center of $\mathfrak{M}$ (resp. $\mathfrak{N}$).

Proof. The case of $\{\mathfrak{M},\alpha\}$. (i) $\Rightarrow$ (ii): It is clear that $C_{\mathfrak{M}\times_\alpha G} \subset \alpha(\mathfrak{M})$. If $\alpha(z) = z \otimes 1$ for $z \in C_{\mathfrak{M}}$, then $\alpha(z)$ commutes with $\mathfrak{M} \times_\alpha G$. and hence $z = 1$.

(ii) $\Rightarrow$ (i) : Since $\overline{\mathfrak{M}}^{\tilde{\alpha}} = \mathfrak{M} \times_\alpha G$ by Theorem II.1.2, it suffices to show that $\overline{\mathfrak{M}}^{\tilde{\alpha}}$ is a factor.

The ergodicity of α on $C_{\mathfrak{M}}$ is equivalent to that of $\tilde{\alpha}$ on $C_{\overline{\mathfrak{M}}}$. Because,

$$\alpha(x) = x \otimes 1 \iff \tilde{\alpha}(x \otimes 1) = x \otimes 1 \otimes 1 .$$

Moreover, $C_{\mathfrak{M}\times_\alpha G} \subset \alpha(\mathfrak{M})$ is equivalent to $C_{\overline{\mathfrak{M}}\times_{\tilde{\alpha}} G} \subset \tilde{\alpha}(\overline{\mathfrak{M}})$ by Corollary 1.6. Since $\tilde{\alpha}$ is dual, $C_{\overline{\mathfrak{M}}\times_{\tilde{\alpha}} G} \subset \tilde{\alpha}(\overline{\mathfrak{M}})$ is equivalent to $C_{\overline{\mathfrak{M}}^{\tilde{\alpha}}} \subset C_{\overline{\mathfrak{M}}}$ by Proposition 3.1. Here we use the above ergodicity of $\tilde{\alpha}$ on $C_{\overline{\mathfrak{M}}}$, we find that $\overline{\mathfrak{M}}^{\tilde{\alpha}}$ is a factor.

The case of $\{\mathfrak{N},\delta\}$. (i) $\Rightarrow$ (ii) : It is clear that $C_{\mathfrak{N}\times_\delta G} \subset \delta(\mathfrak{N})$. If $\delta(z) = z$ $z \otimes 1$ for $z \in C_{\mathfrak{N}}$, then $\delta(z)$ commutes with $\mathfrak{N} \times_\delta G$ and so $z = 1$.

(ii) $\Rightarrow$ (i): It suffices to show that $\overline{\mathfrak{N}}^{\tilde{\delta}}$ is a factor. The ergodicity of δ on $C_{\mathfrak{N}}$ is equivalent to that of $\tilde{\delta}$ on $C_{\overline{\mathfrak{N}}}$. That $C_{\mathfrak{N}\times_\delta G} \subset \delta(\mathfrak{N})$ is equivalent to $C_{\overline{\mathfrak{N}}\times_{\tilde{\delta}} G} \subset \tilde{\delta}(\overline{\mathfrak{N}})$ by Corollary 1.6. Since $\tilde{\delta}$ is dual, $C_{\overline{\mathfrak{N}}\times_{\tilde{\delta}} G} \subset \tilde{\delta}(\overline{\mathfrak{N}})$ is equivalent to $C_{\overline{\mathfrak{N}}^{\tilde{\delta}}} \subset C_{\overline{\mathfrak{N}}}$. The ergodicity of $\tilde{\delta}$ on $C_{\overline{\mathfrak{N}}}$ implies that $\overline{\mathfrak{N}}^{\tilde{\delta}}$ is a factor.

Q.E.D.

Theorem 3.4. Assume that $\mathfrak{M}^\alpha$ (resp. $\mathfrak{N}^\delta$) is properly infinite. If α (resp. δ) is integrable and $C_{\mathfrak{M}\times_\alpha G} \subset \alpha(\mathfrak{M})$(resp. $\Gamma(\delta) = G$), then α (resp. δ) is dominant.

Proof. The case of $\{\mathfrak{M},\alpha\}$: Since $\mathfrak{M}^\alpha$ is properly infinite and α is integrable, we have

$$\{\mathfrak{M},\alpha\} \cong \{\overline{\mathfrak{M}},\bar{\alpha}\} \cong \{\overline{\mathfrak{M}},\tilde{\alpha}\}_e \tag{3.4}$$

for some projection e in $\overline{\mathfrak{M}}^{\tilde{\alpha}}$. Let f be the central support of e in $\overline{\mathfrak{M}}^{\tilde{\alpha}}$. That $C_{\mathfrak{M}\times_\alpha G} \subset \alpha(\mathfrak{M})$ is equivalent to that $C_{\overline{\mathfrak{M}}\times_{\tilde{\alpha}} G} \subset \tilde{\alpha}(\overline{\mathfrak{M}})$ by Corollary 1.6. Since $\tilde{\alpha}$ is dual and f is central, $\tilde{\alpha}^f$ is also dual by Corollary 3.2.. Since $\mathfrak{M}^\alpha$ is properly infinite, $\overline{\mathfrak{M}}_e^{\tilde{\alpha}}$ and $\overline{\mathfrak{M}}_f^{\tilde{\alpha}}$ are also properly infinite. Therefore, $\tilde{\alpha}^f$ is dominant on $\overline{\mathfrak{M}}_f$.

On the other hand, since $\overline{\mathfrak{M}}_f^{\tilde{\alpha}}$ is σ-finite and e is a properly infinite projection with central support f, it follows that $e \sim f$ in $\overline{\mathfrak{M}}_f^{\tilde{\alpha}}$. Therefore

$$\{\overline{\mathfrak{M}},\tilde{\alpha}\}_e \cong \{\overline{\mathfrak{M}},\tilde{\alpha}\}_f ,$$

which implies that α is dominant on $\mathfrak{M}$ by (3.4).

The case of $\{\mathcal{M},\delta\}$: Since $\mathcal{M}^{\delta}$ is properly infinite and δ is integrable, we have

$$(3.5) \qquad \{\mathcal{M},\delta\} \cong \{\overline{\mathcal{M}},\overline{\delta}\} \cong \{\overline{\mathcal{M}},\tilde{\delta}\}_e$$

for some projection e in $\overline{\mathcal{M}}^{\tilde{\delta}}$. The rest of the proof is the same as the above case.

Q.E.D.

NOTES

A necessary and sufficient condition, Theorem 3.3, for a crossed product to be a factor was obtained in terms of $\Gamma(\alpha)$ by Connes-Takesaki [14] for the abelian case and by Nakagami for the non-abelian case [47]. Theorem 3.4 is due to [14] for α and [47] for δ.

§4. Co-actions and Roberts actions.

In this section we shall discuss the relation between a Roberts action and a co-action. We indicate by $\{\pi_r, \mathfrak{H}_r\}$ a representation of G which is quasi-equivalent to the right regular representation of G.

Given a ring $\mathcal{R}$ of representations of G we consider two objects $\pi_{\mathcal{R}}$, $v_{\mathcal{R}}$ and the relation between them. Now we set

$$(4.1) \qquad \{\pi_{\mathcal{R}}, \mathfrak{H}_{\mathcal{R}}\} = \sum_{\pi \in \mathcal{R}}^{\oplus} \{\pi, \mathfrak{H}_{\pi}\} .$$

Then $\pi_{\mathcal{R}}$ is a unitary representation of G on $\mathfrak{H}_{\mathcal{R}}$. Each element of $\mathcal{L}(\mathfrak{H}_{\mathcal{R}})$ is represented by a matrix $(a_{\pi,\pi'})_{\pi,\pi' \in \mathcal{R}}$. Condition (I.3.3) implies that

$$(a_{\pi,\pi'}) \in \pi_{\mathcal{R}}(G)' \iff a_{\pi,\pi'} \in \mathcal{J}_G(\pi,\pi') \quad \text{for all} \quad \pi,\pi' \in \mathcal{R} .$$

For each irreducible representation $\{\pi, \mathfrak{H}_{\pi}\} \in \mathcal{R}$ and for each $\xi \in \mathfrak{H}_{\pi}$ we define an operator $v_{\mathcal{R}}(\xi)$ on $\mathfrak{H}_{\mathcal{R}}$ by

$$(4.2) \qquad v_{\mathcal{R}}(\xi) \sum^{\oplus} \eta_{\pi'} = \sum^{\oplus} \eta_{\pi'} \otimes \xi , \qquad \eta_{\pi'} \in \mathfrak{H}_{\pi'} .$$

Then it satisfies

$$(4.3) \qquad \begin{aligned} \|v_{\mathcal{R}}(\xi)\| &= \|\xi\| , & \xi &\in \mathfrak{H}_{\pi} \\ v_{\mathcal{R}}(\xi \otimes \eta) &= v_{\mathcal{R}}(\eta) v_{\mathcal{R}}(\xi) , & \eta &\in \mathfrak{H}_{\pi'} . \end{aligned}$$

In particular, if $\|\xi\| = 1$, then $v_{\mathcal{R}}(\xi)$ is an isometry. Moreover

$$(4.4) \qquad \pi_{\mathcal{R}}(t) v_{\mathcal{R}}(\xi) = v_{\mathcal{R}}(\pi(t)\xi) \pi_{\mathcal{R}}(t) , \qquad t \in G .$$

Now suppose a Roberts action $\{\rho, \eta\}$ of $\mathcal{R}$ on a von Neumann algebra $\{\mathfrak{h}, \mathfrak{K}\}$ is given. We set

$$\{\rho_{\mathcal{R}}, \mathfrak{K}_{\mathcal{R}}\} = \sum^{\oplus} \{\rho_{\pi}, \mathfrak{K}_{\pi}\} , \qquad (\mathfrak{K}_{\pi} = \mathfrak{K}) ,$$

considering each ρ_{π} as a representation of $\mathfrak{h}$ on $\mathfrak{K}_{\pi} = \mathfrak{K}$. As $\mathcal{R}$ contains trivial representation, $\rho_{\mathcal{R}}$ is an isomorphism of $\mathfrak{h}$ into $\mathfrak{h} \bar{\otimes} \ell^{\infty}(\mathcal{R})$. Condition (I.3.2) implies

$$(b_{\pi,\pi'}) \in \rho_{\mathcal{R}}(\mathfrak{h})' \iff b_{\pi,\pi'} \in \mathcal{J}_{\mathfrak{h}}(\rho_{\pi}, \rho_{\pi'}) \quad \text{for all} \quad \pi,\pi' \in \mathcal{R} .$$

Conditions (iii) and (iv) in the definition of Roberts actions give us a *-homomorphism η of $\pi_{\mathcal{R}}(G)'$ into $\rho_{\mathcal{R}}(\mathfrak{h})'$ such that

(4.5) $$\eta((a_{\pi,\pi'})) = (\eta_{\pi,\pi'}(a_{\pi,\pi'})) \ , \quad a_{\pi,\pi'} \in \mathcal{J}_G(\pi,\pi') \ .$$

If $a_{\pi,\pi'}$ is an isometry, so is $\eta_{\pi,\pi'}(a_{\pi,\pi'})$ since $\eta_{\pi,\pi}(1) = 1$.

Next we want to construct a Hilbert space on which we realize the crossed product of $\mathfrak{n}$ by $\mathfrak{R}$ with respect to the Roberts action $\{\rho,\eta\}$. Let $F(\rho,\mathfrak{R})$ be the set of all bounded linear operators Φ of $\mathfrak{H}_{\mathfrak{R}}$ to $\mathfrak{R}$ satisfying

(4.6) $$\eta_{\pi,\pi'}(a_{\pi,\pi'})\Phi\xi_{\pi'} = \Phi a_{\pi,\pi'}\xi_{\pi'} , \qquad \xi_{\pi'} \in \mathfrak{H}_{\pi'} \ .$$

If we denote the restriction of Φ to $\mathfrak{H}_\pi$ by $\Phi(\pi)$, then it is written by

(4.7) $$\eta_{\pi,\pi'}(a_{\pi,\pi'})\Phi(\pi') = \Phi(\pi)a_{\pi,\pi'} \ .$$

Assume that $\mathfrak{R}$ has an element $\{\pi_r,\mathfrak{H}_r\}$ quasi-equivalent to the right regular representation of G. Then there exists an isomorphism θ of $\pi_r(G)''$ onto $\mathfrak{R}(G)$ with $\theta(\pi_r(t)) = \rho(t)$. Let $F_0(\rho,\mathfrak{R})$ be the set of all $\Phi \in F(\rho,\mathfrak{R})$ with

(4.8) $$\|\Phi\|^2 = \psi_G(\theta(\Phi(\pi_r)^*\Phi(\pi_r))) < \infty \ ,$$

where ψ_G is the Plancherel weight on $\mathfrak{R}(G)$.

<u>Lemma</u> 4.1. If $\Phi(\pi_r) = 0$, then $\Phi = 0$.

<u>Proof</u>. Let θ be the isomorphism of $\pi_r(G)''$ onto $\mathfrak{R}(G)$ with $\theta(\pi_r(t)) = \rho(t)$. Since

$$\Phi(\pi_r \otimes 1)a = \eta_{\pi_r\otimes 1,\pi_r}(a)\Phi(\pi_r) = 0 \ , \qquad a \in \mathcal{J}_G(\pi_r \otimes 1,\pi_r)$$

for any trivial representation 1, we have $\Phi(\pi_r \otimes 1) = 0$. If $\{\pi,\mathfrak{H}_\pi\} \in \mathfrak{R}$ and U is a unitary in $L^\infty(G) \mathbin{\overline{\otimes}} \mathfrak{L}(\mathfrak{H}_\pi)$ with $(U\xi)(t) = \pi(t)\xi(t)$ for $\xi \in L^2(G) \otimes \mathfrak{H}_\pi$, then

(4.9) $$\mathrm{Ad}_{(\theta\otimes\iota)^{-1}(U)}(\pi_r(t) \otimes 1) = \pi_r(t) \otimes \pi(t)$$

and hence $\Phi(\pi_r \otimes \pi) = 0$. Since

$$\Phi(\pi)b = \eta_{\pi,\pi_r\otimes\pi}(b)\Phi(\pi_r \otimes \pi) \ , \qquad b \in \mathcal{J}_G(\pi,\pi_r \otimes \pi) \ ,$$

we have $\Phi(\pi) = 0$. Since $\{\pi,\mathfrak{H}_\pi\}$ is arbitrary in $\mathfrak{R}$, $\Phi = 0$. Q.E.D.

From this lemma, we know that $\| \ \|$ defined by (4.8) is a norm on $F_0(\rho,\mathfrak{R})$. We denote the completion by $L^2(\rho,\mathfrak{R})$.

<u>Lemma</u> 4.2. If $\Phi \in F(\rho,\mathfrak{R})$, then $\rho(y)\Phi$, $\Phi v_{\mathfrak{R}}(\xi) \in F(\rho,\mathfrak{R})$ for $y \in \mathfrak{n}$ and and $\xi \in \mathfrak{H}_\pi$, where

$$(\rho(y)\Phi)(\pi) = \rho_\pi(y)\Phi(\pi) .$$

Proof. Since $\rho_{\mathfrak{R}}(y)\eta(b) = \eta(b)\rho_{\mathfrak{R}}(y)$ for $b \in \pi_{\mathfrak{R}}(G)'$, it follows that, for any $a \in \mathcal{J}_G(\pi,\pi')$ and $\tilde{a} = \eta_{\pi,\pi'}(a)$,

$$\tilde{a}(\rho(y)\Phi)(\pi') = \tilde{a}\rho_{\pi'}(y)\Phi(\pi') = \rho_\pi(y)\tilde{a}\Phi(\pi')$$

$$= \rho_\pi(y)\Phi(\pi)a \quad = (\rho(y)\Phi)(\pi)a .$$

Similarly, since for any $(a\otimes 1_\pi)^\sim = \eta_{\pi''\otimes\pi,\pi'\otimes\pi}(a\otimes 1_\pi)$, $\tilde{a} = \eta_{\pi'',\pi'}(a)$ and $\tilde{1}_\pi = \eta_{\pi,\pi}(1_\pi)$,

$$(a\otimes 1_\pi)^\sim = \tilde{a}\rho_{\pi!}(\tilde{1}_\pi) , \qquad a \in \mathcal{J}_G(\pi'',\pi')$$

by condition (ii) in the definition of Roberts action, it follows that

$$\Phi v_{\mathfrak{R}}(\xi)a\xi_{\pi'} = \Phi(\pi''\otimes\pi)(a\otimes 1_\pi)(\xi_{\pi'}\otimes\xi)$$

$$= (a\otimes 1_\pi)^\sim\Phi(\pi'\otimes\pi)(\xi_{\pi'}\otimes\xi)$$

$$= \tilde{a}\Phi(\pi'\otimes\pi)(\xi_{\pi'}\otimes\xi) = a\Phi v_{\mathfrak{R}}(\xi)\xi_{\pi'} .$$

Q.E.D.

Therefore we can define a crossed product as the following:

Definition 4.3. Assume that $\mathfrak{R}$ has an element quasi-equivalent to the right regular representation of G. Let $\rho(y)$ and $V(\xi)$ be operators on $L^2(\rho,\mathfrak{R})$ defined by

$$(\rho(y)\Phi)(\pi) = \rho_\pi(y)\Phi(\pi), \quad y \in \mathfrak{h} , \tag{4.10}$$

$$V(\xi)\Phi = \Phi v_{\mathfrak{R}}(\xi), \qquad \xi \in \mathfrak{H}_\pi \tag{4.11}$$

for $\Phi \in E_0(\rho,\mathfrak{R})$. The crossed product of $\mathfrak{h}$ by $\mathfrak{R}$ with respect to a Roberts action $\{\rho,\eta\}$ is the von Neumann algebra generated by $\rho(\mathfrak{h})$ and $V(\mathfrak{H}_\pi), \pi \in \mathfrak{R}$, which will be denoted by $\mathfrak{h}\times_\rho \mathfrak{R}$.

It is immediate from (4.10) and (4.11) that

$$V(\xi)\rho(y) = \rho(\rho_\pi(y))V(\xi) , \quad \xi \in \mathfrak{H}_\pi . \tag{4.12}$$

Lemma 4.4. If $\{\pi,\mathfrak{H}_\pi\} \in \mathfrak{R}$ and $\xi,\eta \in \mathfrak{H}_\pi$, then

$$V(\eta)^*V(\xi) = (\xi|\eta)1 . \tag{4.13}$$

Proof. Since π_r is quasi-equivalent to $\pi_r\otimes\pi$ by (4.9), there exists an isomorphism σ of $\pi_r(G)''$ onto $(\pi_r\otimes\pi)(G)''$ with $\sigma(\pi_r(t)) = \pi_r(t)\otimes\pi(t)$. If $a \in \mathcal{J}_G(\pi_r\otimes\pi,\pi_r)$ is an isometry and $e = a\,a^{-1}$, then

$$e(\sigma(\pi_r(t)))e = Ad_a(\pi_r(t)) \ .$$

If $f = \eta_{\pi_r \otimes \pi, \pi_r}(a)^* \eta_{\pi_r \otimes \pi, \pi_r}(a)$, then

$$\begin{aligned} Ad_a(\Phi(\pi_r)^*\Phi(\pi_r)) &= Ad_a(\Phi(\pi_r)^* f \, \Phi(\pi_r)) \\ &= e(\Phi(\pi_r \otimes \pi)^* \Phi(\pi_r \otimes \pi))e \ . \end{aligned}$$

Since a is an arbitrary isometry in $\mathcal{J}_G(\pi_r \otimes \pi, \pi_r)$ and since $\Phi(\pi_r)^*\Phi(\pi_r) \in \pi_r(G)''$ and $\Phi(\pi_r \otimes \pi)^*\Phi(\pi_r \otimes \pi) \in (\pi_r \otimes \pi)(G)''$, it follows that

$$\sigma(\Phi(\pi_r)^*\Phi(\pi_r)) = \Phi(\pi_r \otimes \pi)^*\Phi(\pi_r \otimes \pi) \ .$$

If $\Phi, \Psi \in F_o(\rho, \mathfrak{R})$ and $\xi, \eta \in \mathfrak{H}_\pi$, then

$$\begin{aligned} (V(\xi)\Phi | V(\eta)\Psi) &= \psi_G(\theta(v_{\mathfrak{R}}(\eta)^*\Psi(\pi_r \otimes \pi)^*\Phi(\pi_r \otimes \pi)v_{\mathfrak{R}}(\xi))) \\ &= \psi_G(\theta(v_{\mathfrak{R}}(\eta)^*\sigma(\Psi(\pi_r)^*\Phi(\pi_r))v_{\mathfrak{R}}(\xi))) \ , \end{aligned} \tag{4.14}$$

where θ is an isomorphism of $\pi_r(G)''$ onto $\mathfrak{R}(G)$ with $\theta(\pi_r(t)) = \rho(t)$.

On the other hand, if $f \in A(G)$ and $\zeta \in \mathfrak{H}_r$, then

$$\begin{aligned} v_{\mathfrak{R}}(\eta)^*\sigma(\pi_r(f))v_{\mathfrak{R}}(\xi)\zeta &= v_{\mathfrak{R}}(\eta)^*\sigma(\pi_r(f))(\zeta \otimes \xi) \\ &= v_{\mathfrak{R}}(\eta)^* \int f(t)(\pi_r(t)\zeta \otimes \pi(t)\xi)dt = \int f(t)(\pi(t)\xi|\eta)\pi_r(t)\zeta \, dt \\ &= \pi_r(g)\zeta \ , \end{aligned}$$

where $g(t) = f(t)(\pi(t)\xi|\eta)$. Therefore, the right hand side of (4.14) is equal to $(\xi|\eta)(\Phi|\Psi)$ by (4.8). Thus (4.13) is proved. Q.E.D.

Now we define a unitary representation U of G on $L^2(\rho, \mathfrak{R})$ by

$$U(t)\Phi = \Phi\pi_{\mathfrak{R}}(t)^{-1} \ , \qquad \Phi \in F_o(\rho, \mathfrak{R}) \ . \tag{4.15}$$

Then, it is immediate that

$$\begin{aligned} U(t)\rho(y) &= \rho(y)U(t) \ , & y \in \mathfrak{h} \ , \\ U(t)V(\xi) &= V(\pi(t)\xi)U(t) \ , & \xi \in \mathfrak{H}_\pi \ . \end{aligned} \tag{4.16}$$

Therefore, $\mathfrak{h} \times_\rho \mathfrak{R}$ is globally invariant under the action $t \to Ad_{U(t)}$, whose restriction to $\mathfrak{h} \times_\rho \mathfrak{R}$ is called the <u>dual</u> action of the Roberts action $\{\rho, \eta\}$ and denoted by $\hat{\rho}$.

Lemma 4.5 Let $\hat{\rho}$ be the dual of a Roberts action $\{\rho,\eta\}$ of $\mathcal{R}$ on $\mathfrak{h}$. If $\{\pi,\mathfrak{H}_\pi\} \in \mathcal{R}$, then

(i) $\{\hat{\rho},V(\mathfrak{H}_\pi)\} \in \mathcal{H}_{\hat{\rho}}(\mathfrak{h} \times_\rho \mathcal{R})$.

(ii) $\{\pi,\mathfrak{H}_\pi\} \cong \{\hat{\rho},V(\mathfrak{H}_\pi)\}$.

Proof (i) Let $\{\iota,\mathfrak{H}_\iota\}$ be a trivial representation of G with $\dim \mathfrak{H}_\iota = 1$ and ε_o a normalized vector with $\mathfrak{H}_\iota = C\varepsilon_o$. Let $\{\varepsilon_1,\ldots,\varepsilon_n\}$ and $\{\bar{\varepsilon}_1,\ldots,\bar{\varepsilon}_n\}$ be the corresponding orthonormal bases of $\mathfrak{H}_\pi$ and $\mathfrak{H}_{\bar{\pi}}$, respectively. Take $a \in \mathcal{J}_G(\bar{\pi} \otimes \pi,\iota)$ and $\bar{a} \in \mathcal{J}_G(\pi \otimes \bar{\pi},\iota)$ so that

$$a\varepsilon_o = \sum \bar{\varepsilon}_j \otimes \varepsilon_j \ , \quad \bar{a}\varepsilon_o = \sum \varepsilon_j \otimes \bar{\varepsilon}_j \ .$$

Since $(a\varepsilon_o \mid \bar{\varepsilon}_j \otimes \varepsilon_k) = \delta_{j,k}$, it follows that

$$(1_\pi \otimes a^*)(\bar{a} \otimes 1_\pi)(\varepsilon_o \otimes \varepsilon_k) = (1_\pi \otimes a^*) \sum_j \varepsilon_j \otimes \bar{\varepsilon}_j \otimes \varepsilon_k = \varepsilon_k \otimes \varepsilon_o \ .$$

Using our assumption $\{\pi,\mathfrak{H}_\pi\} = \{\pi \otimes \iota , \mathfrak{H}_\pi \otimes \mathfrak{H}_\iota\} = \{\iota \otimes \pi , \mathfrak{H}_\iota \otimes \mathfrak{H}_\pi\}$, we have $(1_\pi \otimes a^*)(\bar{a} \otimes 1_\pi) = 1_\pi$ and hence

$$\rho_\pi(\eta_{\bar{\pi}\otimes\pi,\iota}(a)^*)\eta_{\pi\otimes\bar{\pi},\iota}(\bar{a}) = 1 \ . \tag{4.17}$$

From the above assumption, $V(\varepsilon_o) = 1$. Since $V(b\xi) = \rho(\eta_{\pi',\pi}(b))V(\xi)$ for $\xi \in \mathfrak{H}_\pi$ and $b \in \mathcal{J}_G(\pi',\pi)$, it follows that

$$\begin{aligned}\rho(\eta_{\bar{\pi}\otimes\pi,\iota}(a)) &= \rho(\eta_{\bar{\pi}\otimes\pi,\iota}(a))V(\varepsilon_o) = V(a\varepsilon_o) \\ &= \sum V(\bar{\varepsilon}_j \otimes \varepsilon_j) = \sum V(\bar{\varepsilon}_j)V(\varepsilon_j) \ ,\end{aligned}$$

and hence, by Lemma 4.4,

$$\begin{aligned}\sum V(\varepsilon_j)V(\varepsilon_j)^* &= \sum V(\varepsilon_j)\rho(\eta_{\bar{\pi}\otimes\pi,\iota}(a))^*V(\bar{\varepsilon}_j) \\ &= \sum \rho(\rho_\pi(\eta_{\bar{\pi}\otimes\pi,\iota}(a))^*)V(\varepsilon_j)V(\bar{\varepsilon}_j) \qquad \text{by (4.12)} \ , \\ &= \rho(\rho_\pi(\eta_{\bar{\pi}\otimes\pi,\iota}(a))^*)V(\bar{a}\varepsilon_o) \\ &= \rho(\rho_\pi(\eta_{\bar{\pi}\otimes\pi,\iota}(a))^*\eta_{\pi\otimes\bar{\pi},\iota}(\bar{a})) = 1 \qquad \text{by (4.17)} \ .\end{aligned}$$

Therefore $V(\mathfrak{H}_\pi)$ is a Hilbert space in $\mathcal{N} \times_\rho \mathcal{R}$. Since $\hat{\rho}_t(V(\xi)) = V(\pi(t)\xi)$ for $\xi \in \mathfrak{H}_\pi$ by (4.16), it follows that $\{\hat{\rho}, V(\mathfrak{H}_\pi)\} \in \mathcal{H}_{\hat{\rho}}(\mathcal{N} \times_\rho \mathcal{R})$.

(ii) It is clear from

$$V(\eta)^*\hat{\rho}_t(V(\xi)) = V(\eta)^*V(\pi(t)\xi) = (\pi(t)\xi|\eta)1$$

for $\xi,\eta \in \mathfrak{H}_\pi$. Q.E.D.

In the rest of this section we assume that G is compact and dt the normalized Haar measure.

<u>Proposition</u> 4.6. $(\mathcal{N} \times_\rho \mathcal{R})^{\hat{\rho}} = \rho(\mathcal{N})$.

<u>Proof</u>. By (4.16) it suffices to show that $(\mathcal{N} \times_\rho \mathcal{R})^{\hat{\rho}} \subset \rho(\mathcal{N})$. First we recall that, if $\{\pi,\mathfrak{H}_\pi\} \in \mathcal{R}$ is irreducible, then

$$\int \pi(t)\xi \; dt = 0 \;, \quad \xi \in \mathfrak{H}_\pi$$

or π is a trivial representation, where dt is the normalized Haar measure. Therefore

$$V\left(\int \pi(t)\xi \; dt\right) \in \mathbf{C} \;.$$

Now, we denote by $\mathcal{E}$ the normal expectation of $\mathcal{N} \times_\rho \mathcal{R}$ onto $(\mathcal{N} \times_\rho \mathcal{R})^{\hat{\rho}}$:

$$\mathcal{E}(x) = \int \hat{\rho}_t(x)dt \;, \quad x \in \mathcal{N} \times_\rho \mathcal{R} \;.$$

The set of elements of the form

$$x = \sum \rho(y_j)V(\xi_j) \qquad y_j \in \mathcal{N} \,,\, \xi_j \in \mathfrak{H}_{\pi_j}$$

is σ-weakly dense in $\mathcal{N} \times_\rho \mathcal{R}$ and

$$\mathcal{E}(x) = \sum \rho(y_j)V\left(\int \pi(t)\xi_j dt\right) \in \rho(\mathcal{N}) \;.$$

Q.E.D.

<u>Lemma</u> 4.7. If $\mathcal{H}_\alpha(\mathcal{M})$ contains a representative of p for all $p \in \hat{G}$ and α is implemented by a unitary representation u of G on $\mathfrak{H}$ such that $\alpha(x) = u(x \otimes 1)u^*$ for $x \in \mathcal{M}$, then $\{\xi \in \mathfrak{H} : u(t)\xi = \xi \,,\, t \in G\}$ is cyclic for $\mathcal{M}$.

<u>Proof</u>. Choose a representative $\{\pi_p,\mathfrak{H}_p\} \in \mathcal{H}_\alpha(\mathcal{M})$ of p for each $p \in \hat{G}$. Let $\{v_1,\dots,v_n\}$ be an orthonormal basis of $\mathfrak{H}_p$. Put

$$E_{\bar{\pi}_p} = n\int \sum_{j=1}^{n} v_j^*\alpha_t(v_j)u(t)dt$$

$$E_o = \int u(t)dt \ .$$

Then $E_{\bar{\pi}_p}$ is the central projection in $u(G)''$ and $E_o\mathfrak{H} = \{\xi \in \mathfrak{H} : u(t)\xi = \xi \ , t \in G\}$.

Since

$$E_{\bar{\pi}_p} = n\sum_j v_j^*\left(\int u(t)dt\right) v_j = n\sum v_j^* E_o v_j \ ,$$

it follows that $E_{\pi_p} \mathfrak{H} \subset (\mathfrak{M}E_o\mathfrak{H})^-$. Since $\sum_{p\in\hat{G}} E_{\pi_p} = 1$, $E_o\mathfrak{H}$ is cyclic for $\mathfrak{M}$. Q.E.D.

We are now ready to prove the equivalence between the crossed product by a co-action and that by a Roberts action.

<u>Theorem</u> 4.8. Assume that G is compact and $\mathfrak{n}$ is properly infinite.

(i) If δ is a co-action of G on $\mathfrak{n}$, there exist a ring $\mathcal{R}$ containing an element quasi-equivalent to the right regular representation of G and a Roberts action $\{\rho,\eta\}$ of $\mathcal{R}$ on $\mathfrak{n}$ such that

$$\{\mathfrak{n} \times_\delta G \, , \hat{\delta}\} \cong \{\mathfrak{n} \times_\rho \mathcal{R} \, , \hat{\rho}\} \ . \tag{4.18}$$

(ii) If $\mathcal{R}$ is a ring containing an element quasi-equivalent to the right regular representation of G and $\{\rho,\eta\}$ is a Roberts action of $\mathcal{R}$ on $\mathfrak{n}$, there exists a co-action δ of G on $\mathfrak{n}$ satisfying (4.18).

<u>Proof</u>. (i) Put $\{\mathfrak{M},\alpha\} = \{\mathfrak{n} \times_\delta G \, , \hat{\delta}\}$. Then $\mathcal{H}_\alpha(\mathfrak{M})$ is a self-adjoint ring. If we define $\{\rho,\eta\}$ by

$$\delta(\rho_\pi(y)) = \rho_{\mathfrak{H}_\pi}(\delta(y)), \quad \mathfrak{H}_\pi \in \mathcal{H}_\alpha(\mathfrak{M}) \ , \tag{4.19}$$

$$\delta(\eta_{\pi,\pi'}(b)) = b \ , \quad b \in \mathcal{J}_G(\mathfrak{H}_\pi,\mathfrak{H}_{\pi'}) \ , \quad \mathfrak{H}_{\pi'} \in \mathcal{H}_\alpha(\mathfrak{M}) \ , \tag{4.20}$$

then it is a Roberts action of $\mathcal{H}_\alpha(\mathfrak{M})$ on $\mathfrak{n}$. Since α is a dual action, there exists a unitary w in $\mathfrak{M} \bar{\otimes} \mathcal{R}(G)$ such that $(\alpha_t \otimes \iota)(w^*) = w^*(1 \otimes \rho(t))$ by Theorem II.2.2. Since $\mathfrak{M}^\alpha$ is properly infinite, there exists an element in $\mathcal{H}_\alpha(\mathfrak{M})$ quasi-equivalent to the right regular representation of G by Proposition 2.3. Therefore we can construct a crossed product $\mathfrak{n} \times_\rho \mathcal{R}$, where $\mathcal{R} = \mathcal{H}_\alpha(\mathfrak{M})$.

Let E_o be the projection of $\mathcal{R} \otimes L^2(G)$ onto $\mathcal{R} \otimes \mathbb{C}1$, where 1 is a constant 1 function. We shall identify $\mathcal{R}$ with $\mathcal{R} \otimes \mathbb{C}1$. Then

$$E_o\delta(y) = \delta(y)E_o = y \ , \qquad y \in \mathfrak{n} \ , \tag{4.21}$$

(4.22) $$E_o b = bE_o = \eta_{\pi',\pi}(b) , \qquad b \in \mathcal{I}_G(\pi',\pi) .$$

Now, for any $\xi \in \mathfrak{K} \otimes L^2(G)$ we define Φ_ξ by

$$\Phi_\xi(\pi)a = E_o a\xi , \quad a \in \mathfrak{H}_\pi , \quad \pi \in \mathfrak{R} .$$

Then Φ_ξ is a bounded linear operator of $\sum^\oplus \mathfrak{H}_\pi$ to $\mathfrak{K}$ and $\Phi_\xi \in F_o(\rho,\mathfrak{K})$. Indeed, if $b \in \mathcal{I}_G(\pi',\pi)$, then

$$\Phi_\xi(\pi')ba = E_o ba\xi = \eta_{\pi',\pi}(b)E_o a\xi = \eta_{\pi',\pi}(b)\Phi_\xi(\pi)a ,$$

by (4.22). Moreover, if $\xi,\eta \in \mathfrak{K}$, $f,g \in L^2(G)$ and $a,b \in \mathfrak{H}_{\pi_r}$, then

$$(\Phi_{\eta\otimes g}(\pi_r)^*\Phi_{\xi\otimes f}(\pi_r)a\,|\,b) = (b^*E_o a(\xi \otimes f)\,|\,\eta \otimes g)$$

$$= \int (b^*(1 \otimes \lambda(t))a(\xi \otimes f)\,|\,\eta \otimes g)dt$$

$$= \int b^*\delta_t(a)(\xi\,|\,\eta)(\lambda(t)f\,|\,g)dt ,$$

and hence

$$\Phi_{\eta\otimes g}(\pi_r)^*\Phi_{\xi\otimes f}(\pi_r) = (\xi\,|\,\eta)\,\pi_r(\varphi^\vee) , \quad \varphi = \omega_{f,g} ,$$

which implies that

$$\|\Phi_{\xi\otimes f}\|^2 = \psi_G(\theta(\Phi_{\xi\otimes f}(\pi_r)^*\Phi_{\xi\otimes f}(\pi_r))) = \|\xi \otimes f\|^2 .$$

Therefore we have an isometry W of $\mathfrak{K} \otimes L^2(G)$ into $L^2(\rho,\mathfrak{K})$ such that $W\xi = \Phi_\xi$ for $\xi \in \mathfrak{K} \otimes L^2(G)$.

Next we shall show by using (4.21) and (4.22) that

(4.23) $$W\delta(y) = \rho(y)W , \qquad y \in \mathfrak{h} ,$$

(4.24) $$Wa = V(a)W , \qquad a \in \mathfrak{H}_\pi , \ \pi \in \mathfrak{R} ,$$

(4.25) $$W(1 \otimes \lambda(t)) = U(t)W , \qquad t \in G .$$

If $\xi \in \mathfrak{K} \otimes L^2(G)$ and $b \in \mathfrak{H}_\pi$, then $\Phi_{\delta(y)\xi}(\pi)b = E_o b\delta(y)\xi = E_o\rho_{\mathfrak{H}_\pi}(\delta(y))b\xi = \rho_{\mathfrak{H}_\pi}(\delta(y))E_o b\xi = \delta(\rho_\pi(y))E_o b\xi = \rho_\pi(y)E_o b\xi = \rho_\pi(y)\Phi_\xi(\pi)b$, and so

$$W\delta(y)\xi = \Phi_{\delta(y)\xi} = \rho(y)\Phi_\xi = \rho(y)W\xi .$$

If $\xi \in \mathfrak{R} \otimes L^2(G)$ and $b \in \mathfrak{H}_{\pi'}$, then $\Phi_\xi v_{\mathfrak{R}}(a)b = \Phi_\xi ba = E_o ba\xi = \Phi_{a\xi}(\pi')b$, and so

$$Wa\xi = \Phi_{a\xi} = \Phi_\xi v_{\mathfrak{R}}(a) = V(a)\Phi_\xi = V(a)W\xi .$$

If $\xi \in \mathfrak{R} \otimes L^2(G)$ and $b \in \mathfrak{H}_\pi$, then $\Phi_{(1\otimes\lambda(t))\xi}(\pi)b = E_o b(1 \otimes \lambda(t))\xi = E_o\hat{\delta}_t^{-1}(b)\xi = \Phi_\xi\hat{\delta}_t^{-1}(b) = (\Phi_\xi\pi_{\mathfrak{R}}(t)^{-1})(\pi)b$, and so

$$W(1 \otimes \lambda(t))\xi = \Phi_{(1\otimes\lambda(t))\xi} = \Phi_\xi\pi_{\mathfrak{R}}(t)^{-1} = U(t)W\xi .$$

Using (4.23), (4.24) and (4.25), we shall show that W is an isometry of $\mathfrak{R} \otimes L^2(G)$ onto $L^2(\rho,\mathfrak{R})$. If $\Phi \in L^2(\rho,\mathfrak{R})$ and $\Phi = U(t)\Phi$ $(= \Phi\pi_{\mathfrak{R}}(t)^{-1})$ for all $t \in G$, then $\Phi(\pi) = 0$ for all irreducible and non trivial $\{\pi,\mathfrak{H}_\pi\} \in \mathfrak{R}$. If $\{\pi,\mathfrak{H}_\pi\}$ is a trivial representation, then $\mathfrak{H}_\pi \subset \delta(\mathfrak{n})$ and hence $\rho_\pi(y) = u_\pi y u_\pi^*$ for some unitary $u_\pi \in \mathfrak{n}$ by (4.19). If $\{\pi,\mathfrak{H}_\pi\}$ and $\{\pi',\mathfrak{H}_{\pi'}\}$ are trivial, that is, $\mathfrak{H}_\pi = C\delta(u_\pi)$ and $\mathfrak{H}_{\pi'} = C\delta(u_{\pi'})$, then $a \in \mathcal{I}_G(\pi',\pi)$ is of the form $\mu\delta(u_{\pi'}u_\pi^*)$ for some non zero $\mu \in C$ and $\eta_{\pi',\pi}(a) = \mu u_{\pi'}u_\pi^*$ by (4.20). Here we set $\xi_\pi = \Phi(\pi)\delta(u_\pi)$ for a trivial π. Then $\Phi = \Phi_{u_\pi^*\xi_\pi\otimes 1}$. Indeed, if π and π' are trivial, then

$$\Phi_{u_\pi^*\xi_\pi\otimes 1}(\pi)\delta(u_\pi) = E_o\delta(u_\pi)(u_\pi^*\xi_\pi \otimes 1) = \xi_\pi = \Phi(\pi)\delta(u_\pi)$$

and

$$\begin{aligned}\Phi_{u_\pi^*\xi_\pi\otimes 1}(\pi')\delta(u_{\pi'}) &= u_{\pi'}u_\pi^*\xi_\pi = \mu^{-1}\eta_{\pi',\pi}(a)\xi_\pi \\ &= \mu^{-1}\eta_{\pi',\pi}(a)\Phi(\pi)\delta(u_\pi) = \mu^{-1}\Phi(\pi')a\delta(u_\pi) = \Phi(\pi')\delta(u_{\pi'}) .\end{aligned}$$

Consequently, we have $\Phi \in WE_o\mathfrak{H}$ $(\mathfrak{H} = \mathfrak{R} \otimes L^2(G))$ and so

$$WE_o\mathfrak{H} = \{\Phi \in L^2(\rho,\mathfrak{R}) : U(t)\Phi = \Phi, t \in G\} . \tag{4.26}$$

Since $\mathfrak{R}$ contains an element quasi-equivalent to the right regular representation of G, so does $\mathcal{H}_{\hat{\rho}}(\mathfrak{n} \times_\rho \mathfrak{R})$ by Lemma 4.5. Therefore $\mathcal{H}_{\hat{\rho}}(\mathfrak{n} \times_\rho \mathfrak{R})$ contains a representative of p for all $p \in \hat{G}$ by Proposition 2.3 and Lemma 2.7. Thus, the right hand side of (4.26) is cyclic for $\mathfrak{n} \times_\rho \mathfrak{R}$ by Lemma 4.7. This means that W is an isometry of $\mathfrak{R} \otimes L^2(G)$ onto $L^2(\rho,\mathfrak{R})$.

Finally, if we notice that $\mathfrak{n} \times_\delta G$ is generated by $\delta(\mathfrak{n})$ and $\{\mathfrak{H}_\pi$: irreducible $\pi \in \mathfrak{R}\}$ by Proposition 2.6 and $\mathfrak{n} \times_\rho \mathfrak{R}$ is generated by $\rho(\mathfrak{n})$ and $\{V(\mathfrak{H}_\pi)$: irreducible $\pi \in \mathfrak{R}\}$, then Ad_W gives the equivalence (4.18).

(ii) Since $\mathfrak{R}$ has an element quasi-equivalent to the right regular representation, we can consider the crossed product $\mathfrak{n} \times_\rho \mathfrak{R}$. Here we set $\{\mathfrak{m},\alpha\} = \{\mathfrak{n} \times_\rho \mathfrak{R}, \hat{\rho}\}$. By Lemma 4.5, $\mathcal{H}_\alpha(\mathfrak{m})$ has also an element quasi-equivalent to the right regular representation. Since $\mathfrak{n}$ is properly infinite, there exists a unitary $w \in \mathfrak{m} \bar{\otimes} \mathcal{L}(L^2(G))$ with $(\alpha_t \otimes \iota)(w^*) = w^*(1 \otimes \rho(t))$ by Proposition 2.3. Therefore, α is a dual action, for $\{\mathfrak{m},\alpha\} \cong \{\bar{\mathfrak{m}},\bar{\alpha}\} \cong \{\bar{\mathfrak{m}},\tilde{\alpha}\}$. Q.E.D.

From this theorem we know that if we consider the following correspondences:

$$\{\mathfrak{n},G,\delta_1\} \xrightarrow{(i)} \{\mathfrak{n},\mathfrak{K},\{\rho,\eta\}\} \xrightarrow{(ii)} \{\mathfrak{n},G,\delta_2\} ,$$

then $\{\mathfrak{n} \times_{\delta_1} G, \hat{\delta}_1\} \cong \{\mathfrak{n} \times_{\delta_2} G, \hat{\delta}_2\}$. Namely, there is an isomorphism $\bar{\pi}$ of $\mathfrak{n} \times_{\delta_1} G$ onto $\mathfrak{n} \times_{\delta_2} G$ with $(\bar{\pi} \otimes \iota) \circ \hat{\delta}_1 = \hat{\delta}_2 \circ \bar{\pi}$. Since $1 \otimes W_G \in (\mathfrak{n} \times_{\delta_j} G) \bar{\otimes} \mathcal{R}(G)$, we may set $w_1 = (\bar{\pi} \otimes \iota)(1 \otimes W_G)$ and $w_2 = 1 \otimes W_G$. Then $((\delta_2)^{\wedge}_t \otimes \iota)(w_2^*) = w_2^*(1 \otimes \rho(t))$. Since $(\delta_2)^{\wedge}_t \circ \bar{\pi} = \bar{\pi} \circ (\delta_1)^{\wedge}_t$, we have $((\delta_1)^{\wedge}_t \otimes \iota)(w_1^*) = w_1^*(1 \otimes \rho(t))$. It follows that $w_1^* w_2 \in \delta_2(\mathfrak{n}) \bar{\otimes} \mathcal{R}(G)$ and $(w_1^* w_2 \otimes 1)(\iota \otimes \delta_G \otimes \iota)(w_1^* w_2) = (\iota \otimes \iota \otimes \delta_G)(w_1^* w_2)$. Therefore $\{\mathfrak{n},\delta_1\}$ and $\{\mathfrak{n},\delta_2\}$ are outer conjugate:

$$\{\mathfrak{n},\delta_1\} \cong \{\delta_1(\mathfrak{n}), \iota \otimes \delta_G\} \cong \{\delta_2(\mathfrak{n}), \mathrm{Ad}_{w_1^*}\} \sim \{\delta_2(\mathfrak{n}), \iota \otimes \delta_G\} \cong \{\mathfrak{n},\delta_2\} .$$

In the same way, if we consider

$$\{\mathfrak{n},\mathfrak{K}_1,\{\rho_1,\eta_1\}\} \xrightarrow{(ii)} \{\mathfrak{n},G,\delta\} \xrightarrow{(i)} \{\mathfrak{n},\mathfrak{K}_2,\{\rho_2,\eta_2\}\} ,$$

then $\{\mathfrak{n} \times_{\rho_1} \mathfrak{K}_1, \hat{\rho}_1\} \cong \{\mathfrak{n} \times_{\rho_2} \mathfrak{K}_2, \hat{\rho}_2\}$. Since $\rho_1(\mathfrak{n})$ corresponds to $\rho_2(\mathfrak{n})$ and $\mathfrak{K}_1$ is realized in $\mathfrak{K}_2$, the covariance equivalence of $\{\mathfrak{n},\{\rho_1,\eta_1\}\}$ and $\{\mathfrak{n},\{\rho_2,\eta_2\}\}$ in some sense will be obtained.

NOTES

The materials of this section are adapted from [55].

CHAPTER V.

PERTURBATIONS OF ACTIONS AND CO-ACTIONS.

Introduction. This chapter is concerned with the non-commutative version of the flow of weights of Connes-Takesaki [14]. Since the "dual" of a non-commutative group is not a group, an action of a non-commutative G does not give rise to an action of a non-commutative group on the cohomology space, even though the comparison theory of 1-cocycles goes well. Contrary to this, the dual of the "dual" of G is the given G itself as the Tannaka-Stinespring-Tatsuuma duality theorem says; thus a co-action δ of G on $\mathfrak{h}$ gives rise to an action of G itself on the cohomology space, i.e. an action on an abelian von Neumann algebra, which is isomorphic to the restriction of the dual action $\hat{\delta}$ to the center $C_{\mathfrak{h}\times_\delta G}$ of the crossed product $\mathfrak{h}\times_\delta G$. The discussion here is adapted from the flow of weights, [14], without major change. Of special interest would be Theorem 2.6 which says that the normalized Connes spectrum $\Gamma_n(\delta)$ is invariant under perturbation by 1-cocycles; hence it is precisely the kernel of the action of G on the cohomology space.

§1. Comparison of 1-cocycles of action and co-action.

In this section we consider the set of all perturbed actions by 1-cocycles of a fixed action α . The dual version for a co-action is also discussed.

Given a strongly continuous map : $t \in G \mapsto v(t) \in \mathfrak{M}$ with partial isometry values satisfying

$$v(s)\,\alpha_s(v(t)) = v(st)$$
(1.1)
$$v(s^{-1}) = \alpha_s^{-1}(v(s)^*) ,$$

we define an operator a in $\mathfrak{M} \bar{\otimes} L^\infty(G)$ by $(a\xi)(t) = v(t)\xi(t)$ for $\xi \in \mathfrak{H} \otimes L^2(G)$. Then

$$(a \otimes 1)(\alpha \otimes \iota)(a) = (\iota \otimes \alpha_G)(a) ,$$
(1.2)
$$aa^* = e_a \otimes 1 , \; a^*a = \alpha(e_a) ,$$

where e_a is the left support of $v(t)$.

Proposition 1.1. If a is a partial isometry in $\mathfrak{M} \bar{\otimes} L^\infty(G)$ satisfying (1.2), there exists a strongly continuous map: $t \in G \to v(t) \in \mathfrak{M}$ satisfying (1.1).

Proof. By (1.2) we have

$$(\alpha \otimes \iota)(a) = (a^* \otimes 1)(1 \otimes V_G)(a \otimes 1)(1 \otimes V_G)^*$$

and hence $\alpha(a(t)) = a^*(\iota \otimes \rho_t)(a)$ locally almost everywhere. Put $v(t) = \alpha^{-1}(a^*(\iota \otimes \rho_t)(a))$. Then $t \to v(t)$ is a strongly continuous map with (1.1).

Q.E.D.

Definition 1.2. (1-cocycles) Let $Z_\alpha(G,\mathfrak{M})$ be the set of all partial isometries a in $\mathfrak{M} \bar{\otimes} L^\infty(G)$ satisfying (1.2). Let $Z_\delta(G,\mathfrak{N})$ be the set of all partial isometries b in $\mathfrak{N} \bar{\otimes} \mathfrak{R}(G)$ such that

$$(b \otimes 1)(\delta \otimes \iota)(b) = (\iota \otimes \delta_G)(b);$$

$$bb^* = e_b \otimes 1 , \quad b^*b = \delta(e_b)$$

for some projection $e_b \in \mathfrak{N}$.

Since $a(t)\alpha_t(e_a)a(t)^* = e_a$, we consider the action ${}_a\alpha$ of G on the reduced von Neumann algebra $\mathfrak{M}_{e_a}$ defined by

$${}_a\alpha(x) = \mathrm{Ad}_a \circ \alpha(x) , \qquad x \in \mathfrak{M}_{e_a} .$$

Similarly, we consider the co-action ${}_b\delta$ of G on $\mathfrak{N}_{e_b}$ defined by

$$_b\delta(y) = \mathrm{Ad}_b \circ \delta(y) , \qquad y \in \mathfrak{n}_{e_b} .$$

We denote the fixed point subalgebras of $\mathfrak{m}_{e_a}$ and $\mathfrak{n}_{e_b}$ with respect to $_a\alpha$ and $_b\delta$ by $\mathfrak{m}^a$ and $\mathfrak{n}^b$, respectively.

Next we shall consider the cohomologous class of 1-cocycles.

<u>Definition</u> 1.3. (Cohomologous) We say that a and b in $Z_\alpha(G,\mathfrak{m})$ (resp. $Z_\delta(G,\mathfrak{n})$) are <u>equivalent</u> and write $a \simeq b$, if there exists an element c in $\mathfrak{m}$(resp. $\mathfrak{n}$) such that

$$b = (c^* \otimes 1)a\alpha(c) , \qquad a = (c \otimes 1)b\alpha(c^*)$$

$$(\text{resp. } b = (c^* \otimes 1)a\delta(c), \quad a = (c \otimes 1)b\delta(c^*)).$$

Moreover, $a \prec b$ if $a \simeq (e \otimes 1)b$ for some projection e in $\mathfrak{m}^b$ (resp. $\mathfrak{n}^b$).

The following proposition is crucial for this section, which deduces the comparison problem of 1-cocycles into the comparison problem of projections in the sense of Murray and von Neumann.

<u>Proposition</u> 1.4. (a) Let $\bar{\mathfrak{m}} = \mathfrak{m} \otimes F_2$ and $\bar{\alpha}_t = \alpha_t \otimes \iota$. For any $a,b \in Z_\alpha(G,\mathfrak{m})$ we define $c \in Z_{\bar{\alpha}}(G,\bar{\mathfrak{m}})$ by

$$c(s) = a(s) \otimes e_{11} + b(s) \otimes e_{22} . \tag{1.3}$$

Then $a \prec b$ (resp. $a \simeq b$) is equivalent to

$$e_a \otimes e_{11} \precsim e_b \otimes e_{22} \qquad (\text{resp. } e_a \otimes e_{11} \sim e_b \otimes e_{22}) \tag{1.4}$$

in $\bar{\mathfrak{m}}^c$.

(b) Let $\bar{\mathfrak{n}} = \mathfrak{n} \otimes F_2$ and $\bar{\delta} = (\iota \otimes \sigma) \circ (\delta \otimes \iota)$. For any $a,b \in Z_\delta(G,\mathfrak{n})$ we define $c \in Z_{\bar{\delta}}(G,\bar{\mathfrak{n}})$ by

$$\bar{c} = (\iota \otimes \sigma)(a \otimes e_{11} + b \otimes e_{22}) . \tag{1.5}$$

Then $a \prec b$ (resp. $a \simeq b$) is equivalent to (1.4) in $\bar{\mathfrak{n}}^c$.

Therefore, if $e_a \otimes e_{11}$ and $e_b \otimes e_{22}$ have the same central support ($= e_a \otimes e_{11} + e_b \otimes e_{22}$) in $\bar{\mathfrak{m}}^c$(resp. $\bar{\mathfrak{n}}^c$) and $\mathfrak{m}^a$ (resp. $\mathfrak{n}^a$) and $\mathfrak{m}^b$(resp. $\mathfrak{n}^b$) are properly infinite, then $a \simeq b$.

Making use of Proposition 1.4, we define a projection $c_a(b)$ in the center of $\mathfrak{m}^\alpha$(resp. $\mathfrak{n}^\delta$) by $c_a(b) \otimes e_{11} =$ (Central support of $e_b \otimes e_{22}$ in $\bar{\mathfrak{m}}^c$)($e_a \otimes e_{11}$). If $\{a_j\}_{j=1}^\infty$ is a sequence in $Z_\alpha(G,\mathfrak{m})$ (resp. $Z_\delta(G,\mathfrak{n})$) such that $\{e_{a_j}\}$ is mutually orthogonal, then Σa_j is also a 1-cocycle ,

$$c_{\Sigma a_j}(b) = \sum c_{a_j}(b) \quad \text{and} \quad c_b(\Sigma a_j) = \vee c_b(a_j) .$$

Let $\{v_j : j \in \mathbb{N}\}$ be an orthonormal basis of a Hilbert space in $\mathfrak{m}$ (resp. $\mathfrak{n}$). If a is a 1-cocycle, then

$$(1.6) \qquad \check{a} = \sum_{j=1}^{\infty} (v_j \otimes 1) a\, \alpha(v_j)^*$$

is also a 1-cocycle such that $a \prec \check{a}$ and $\mathfrak{m}^{\check{a}}$ (resp. $\mathfrak{n}^{\check{a}}$) is properly infinite.

<u>Proposition</u> 1.5. The map c_a is an ordered isomorphism of the set of all equivalence classes of $b \in Z_\delta(G,\mathfrak{n})$ of infinite multiplicity with $b \prec \check{a}$ onto the set of all projections in the center of $\mathfrak{n}^a$, where the infinite multiplicity of b is defined by the proper infiniteness of $\mathfrak{n}^b$.

§2. Dominant 1-cocycles

In the following we call an element a in $Z_\alpha(G,\mathfrak{m})$ (resp. $Z_\delta(G,\mathfrak{n})$) integrable, semi-dual, dual or dominant according as ${}_a\alpha$ (resp. ${}_a\delta$) has the corresponding property.

The following proposition is a restatement of the definition of a semi-dual action and co-action.

Proposition 2.1. Let $\bar{\mathfrak{m}} = \mathfrak{m} \bar{\otimes} \mathfrak{L}(L^2(G))$ (resp. $\bar{\mathfrak{n}} = \mathfrak{n} \bar{\otimes} \mathfrak{L}(L^2(G))$) and $\bar{\alpha} = (\iota \otimes \sigma)\cdot(\alpha \otimes \iota)$ (resp. $\bar{\delta} = (\iota \otimes \sigma)\circ(\delta \otimes \iota)$).

(a) If a unitary $a \in Z_\alpha(G,\mathfrak{m})$ is a semi-dual, then

$$(\iota \otimes \sigma)(a \otimes 1) \simeq (1 \otimes V_G')(\iota \otimes \sigma)(a \otimes 1) \tag{2.1}$$

in $Z_{\bar{\alpha}}(G,\bar{\mathfrak{m}})$.

(b) If a unitary $b \in Z_\delta(G,\mathfrak{n})$ is semi-dual, then

$$(\iota \otimes \sigma)(b \otimes 1) \simeq (1 \otimes W_G)(\iota \otimes \sigma)(b \otimes 1) \tag{2.2}$$

in $Z_{\bar{\delta}}(G,\bar{\mathfrak{n}})$.

Proof. (a) If $a \in Z_\alpha(G,\mathfrak{m})$, then $(\iota \otimes \sigma)(a \otimes 1) \in Z_{\bar{\alpha}}(G,\bar{\mathfrak{m}})$. Since

$$(V_G' \otimes 1)(\sigma \otimes \iota)(1 \otimes V_G') = \mathrm{Ad}_{1 \otimes V_G}(V_G' \otimes 1), \tag{2.3}$$

we have $1 \otimes V_G' \in Z_{\bar{\alpha}}(G,\bar{\mathfrak{m}})$ As $a \otimes 1 \otimes 1$ commutes with $(\alpha \otimes \iota \otimes \iota)(1 \otimes V_G')$, it follows that $(1 \otimes V_G')(\iota \otimes \sigma)(a \otimes 1) \in Z_{\bar{\alpha}}(G,\bar{\mathfrak{m}})$. Since a is semi-dual, there exists a unitary v in $\mathfrak{m} \bar{\otimes} \mathfrak{R}(G)'$ such that $(1 \otimes V_G')({}_a\alpha)^{-}(v^*) = v^* \otimes 1$. Therefore

$$(v \otimes 1)((1 \otimes V_G')(\iota \otimes \sigma)(a \otimes 1))\bar{\alpha}(v^*) = (\iota \otimes \sigma)(a \otimes 1)$$

and hence (2.1) follows.

(b) If $b \in Z_\delta(G,\mathfrak{n})$, then $(\iota \otimes \sigma)(b \otimes 1) \in Z_{\bar{\delta}}(G,\bar{\mathfrak{n}})$. Since

$$(W_G \otimes 1)(\sigma \otimes \iota)(1 \otimes W_G) = \mathrm{Ad}_{1 \otimes W_G^*}(W_G \otimes 1), \tag{2.4}$$

we have $1 \otimes W_G \in Z_{\bar{\delta}}(G,\bar{\mathfrak{n}})$. As $b \otimes 1 \otimes 1$ commutes with $(\delta \otimes \iota \otimes \iota)(1 \otimes W_G)$, it follows that $(1 \otimes W_G)(\iota \otimes \sigma)(b \otimes 1) \in Z_{\bar{\delta}}(G,\bar{\mathfrak{n}})$. Since b is semi-dual, there exists a unitary u in $\mathfrak{n} \bar{\otimes} L^\infty(G)'$ such that $(1 \otimes W_G)({}_b\delta)^{-}(u^*) = u^* \otimes 1$. Therefore

$$(u \otimes 1)((1 \otimes W_G)(\iota \otimes \sigma)(b \otimes 1))\bar{\delta}(u^*) = (\iota \otimes \sigma)(b \otimes 1)$$

and so (2.2) follows. Q.E.D.

Proposition 2.2. (a) If a is a unitary in $Z_\alpha(G,\mathfrak{m})$, then

$$1 \otimes V'_G \simeq (1 \otimes V'_G)(\iota \otimes \sigma)(a \otimes 1) \tag{2.5}$$

in $Z_{\bar\alpha}(G,\bar{\mathfrak{m}})$.

(b) If b is a unitary in $Z_\delta(G,\mathfrak{n})$, then

$$1 \otimes W_G \simeq (1 \otimes W_G)(\iota \otimes \sigma)(b \otimes 1) \tag{2.6}$$

in $Z_{\bar\delta}(G, \bar{\mathfrak{n}})$.

Proof. (a) Since $1 \otimes V'_G \in Z_{\bar\alpha}(G,\bar{\mathfrak{m}})$ and $(1 \otimes V'_G)(\iota \otimes \sigma)(a \otimes 1) \in Z_{\bar\alpha}(G,\bar{\mathfrak{m}})$ by Proposition 2.1, it remains to show (2.5). By direct computation

$$\mathrm{Ad}_{1 \otimes V'^*}(a \otimes 1) = (\iota \otimes \sigma) \circ (\iota \otimes \alpha_G)(a).$$

Since $a \in Z_\alpha(G,\mathfrak{m})$, it follows that

$$\begin{aligned}&(a^* \otimes 1)((1 \otimes V'_G)(\iota \otimes \sigma)(a \otimes 1))\bar\alpha(a)\\ &= (a^* \otimes 1)(1 \otimes V'_G)((\iota \otimes \sigma) \circ (\iota \otimes \alpha_G)(a)) \qquad \text{by (1.2)},\\ &= 1 \otimes V'_G .\end{aligned}$$

(b) Since $1 \otimes W_G \in Z_{\bar\delta}(G,\bar{\mathfrak{n}})$ and $(1 \otimes W_G)(\iota \otimes \sigma)(b \otimes 1) \in Z_{\bar\delta}(G,\bar{\mathfrak{n}})$ by Proposition 2.1, it remains to show (2.6). By direct computation,

$$\mathrm{Ad}_{1 \otimes W_G^*}(b \otimes 1) = (\iota \otimes \sigma) \circ (\iota \otimes \delta_G)(b).$$

Since $b \in Z_\delta(G,\mathfrak{n})$, it follows that

$$(b^* \otimes 1)((1 \otimes W_G)(\iota \otimes \sigma)(b \otimes 1))\bar\delta(b) = 1 \otimes W_G.$$

Q.E.D.

This proposition shows that

$$({}_a\alpha)^\sim \circ \mathrm{Ad}_a = \mathrm{Ad}_{a \otimes 1} \circ \tilde\alpha \quad \text{on } \bar{\mathfrak{m}}, \tag{2.7}$$

$$({}_b\delta)^\sim \circ \mathrm{Ad}_b = \mathrm{Ad}_{b \otimes 1} \circ \tilde\delta \quad \text{on } \bar{\mathfrak{n}}. \tag{2.8}$$

Proposition 2.3 Let a and b be unitaries in $Z_\alpha(G,\mathfrak{m})$ (resp. $Z_\delta(G,\mathfrak{n})$). If a and b are dominant, then $a \simeq b$.

Proof. (a) Since a and b are dominant, they are semi-dual. Combining Propositions 2.1 and 2.2, we find that

$$1 \otimes V_G' \simeq (\iota \otimes \sigma)(a \otimes 1)$$

in $Z_{\bar{\alpha}}(G,\bar{\mathfrak{m}})$. Therefore $\bar{a} = (\iota \otimes \sigma)(a \otimes 1)$ and $\bar{b} = (\iota \otimes \sigma)(b \otimes 1)$ are equivalent in $Z_{\bar{\alpha}}(G,\bar{\mathfrak{m}})$. Therefore, by Proposition 1.4,

$$e_{\bar{a}} \otimes e_{11} \sim e_{\bar{b}} \otimes e_{22} \quad \text{in } \rho^c ,$$

where $\rho = \bar{\mathfrak{m}} \otimes F_2$ and c is given by (1.3) for $\bar{a}$ and $\bar{b}$. Since $\mathfrak{m}^a$ and $\mathfrak{m}^b$ are properly infinite, we have

$$e_a \otimes e_{11} \sim e_{\bar{a}} \otimes e_{11} \sim e_{\bar{b}} \otimes e_{22} \sim e_b \otimes e_{22} .$$

Thus, Proposition 1.4 implies $a \simeq b$.

(b) By the same argument, $e_{\bar{a}} \otimes e_{11} \sim e_{\bar{b}} \otimes e_{22}$ in ρ^c, where $\rho = \bar{\mathfrak{n}} \otimes F_2$ and c is given by (1.5) for $\bar{a}$ and $\bar{b}$. Since $\mathfrak{n}^a$ and $\mathfrak{n}^b$ are properly infinite, $a \simeq b$. Q.E.D.

Theorem 2.4 Let $\mathfrak{m}$ (resp. $\mathfrak{n}$) be propoerly infinite.

(i) There is a dominant unitary a in $Z_\alpha(G,\mathfrak{m})$ (resp. $Z_\delta(G,\mathfrak{n})$), which is unique up to equivalence.

(ii) An element $b \in Z_\alpha(G,\mathfrak{m})$ (resp. $Z_\delta(G,\mathfrak{n})$) is integrable if and only if $b \prec a$.

Proof. (i) The case of $\{\mathfrak{m},\alpha\}$: If $\mathfrak{m}$ is properly infinite, we can choose an orthonormal basis $\{v_j : j \in \mathbb{N}\}$ of a Hilbert space in $\mathfrak{m}$. Put $a = \check{1}$ $(= \Sigma(v_j \otimes 1)\alpha(v_j)^*)$. Then a is a 1-cocycle unitary and $\mathfrak{m}^a$ is properly infinite. Therefore we may assume that $\mathfrak{m}^\alpha$ is properly infinite. Therefore $\{\mathfrak{m},\alpha\}$ is spatially isomorphic to $\{\bar{\mathfrak{m}},\bar{\alpha}\}$. Since $1 \otimes V_G' \in Z_{\bar{\alpha}}(G,\bar{\mathfrak{m}})$ and $\tilde{\alpha} = \mathrm{Ad}_{1\otimes V_G'} \circ \bar{\alpha}$, there exists a dominant unitary in $Z_\alpha(G,\mathfrak{m})$. By Proposition 2.3 it is unique up to equivalence.

The case of $\{\mathfrak{n},\delta\}$: By using the same device as above we may assume that $\mathfrak{n}^\delta$ is properly infinite. Since $1 \otimes W_G \in Z_{\bar{\delta}}(G,\bar{\mathfrak{n}})$ and $\tilde{\delta} = \mathrm{Ad}_{1\otimes W_G} \circ \bar{\delta}$, there exists a dominant unitary in $Z_\delta(G,\mathfrak{n})$. By Proposition 2.3 it is unique up to equivalence.

(ii) It is immediate from Theorems III.3.1 and III.3.2, for we may assume that $\mathfrak{M}^{\alpha}$ (resp. $\mathfrak{N}^{\delta}$) is properly infinite by the same device as above. Q.E.D.

By virtue of Theorem 2.4 and the example in §IV.1, $\Gamma(\delta)$ is not stable under 1-cocycle perturbation, contrast to the abelian case. This defect of the Connes spectrum $\Gamma(\delta)$ is restored as follows:

Definition 2.5. The normalized Connes spectrum $\Gamma_n(\delta)$ is the intersection $\bigcap_{t\in G} t\Gamma(\delta)t^{-1}$ which is the largest normal subgroup contained in $\Gamma(\delta)$.

Theorem 2.6. The normalized Connes spectrum $\Gamma_n(\delta)$ of a co-action δ is invariant under the perturbation by any unitary 1-cocycle in $Z_{\delta}(G,\mathfrak{N})$ and is equal to $\Gamma(\hat{\hat{\delta}})$.

Proof. By our favorite 2×2-matrix arguments, our assertion is equivalent to the claim that if two projections e and f in $\mathfrak{N}^{\delta}$ are equivalent in $\mathfrak{N}$, then $\Gamma_n(\delta^e) = \Gamma_n(\delta^f)$. Let u be a partial isometry in $\mathfrak{N}$ with $u^*u = e$ and $uu^* = f$. Take a non-zero projection $f_1 \in \mathfrak{N}^{\delta}$ with $f_1 \le f$. Let t be a point in $sp_{\delta}(f_1u)$ and $s \in \Gamma_n(\delta^e)$. Let U be an arbitrary compact neighborhood of s. Let K and V be compact neighborhoods of t and $t^{-1}st$ respectively such that $KVK^{-1} \subset U$. Let φ be a function in $A(G)$ such that $\varphi(t) = 1$ and $\mathrm{supp}\,\varphi \subset K$. Set $a = \delta_{\varphi}(f_1u) \neq 0$. We then have $a^*a \in \mathfrak{N}_e$ and $aa^* \in \mathfrak{N}_{f_1}$. Set $e_1 = \bigvee_{\psi\in A(G)} \mathrm{supp}(\delta_{\psi}(a)^*\delta_{\psi}(a))$.
By Lemma 2.7 below, e_1 belongs to $\mathfrak{N}^{\delta^e}$. Since $t^{-1}st \in \Gamma(\delta^{e_1})$,

$$\bigvee\{\mathrm{supp}(x^*x) : x \in \mathfrak{N}^{\delta^{e_1}}(V)\} = \bigvee\{\mathrm{supp}(xx^*) : x \in \mathfrak{N}^{\delta^{e_1}}(V)\} = e_1 ,$$

so that we can find an $x \in \mathfrak{N}^{\delta^{e_1}}(V)$, $\psi_1, \psi_2 \in A(G)$ such that

$$0 \neq \delta_{\psi_1}(a)\, x\, \delta_{\psi_2}(a)^* \in \mathfrak{N}^{\delta^{f_1}}(KVK^{-1}) \subset \mathfrak{N}^{\delta^{f_1}}(U) .$$

Hence $U \cap sp(\delta^{f_1}) \neq \emptyset$; so $s \in sp(\delta^{f_1})$ for every non-zero projection $f_1 \in \mathfrak{N}^{\delta^f}$, that is, $s \in \Gamma(\delta^f)$. Thus $\Gamma_n(\delta^e) \subset \Gamma(\delta^f)$; so $\Gamma_n(\delta^e) \subset \Gamma_n(\delta^f)$ by the normality of $\Gamma_n(\delta^e)$. By symmetry, we have $\Gamma_n(\delta^e) = \Gamma_n(\delta^f)$. Q.E.D.

Lemma 2.7. For any $x \in \mathfrak{N}$, we have

$$\bigvee\{\mathrm{supp}(\delta_{\varphi}(x)^*\delta_{\varphi}(x)) : \varphi \in A(G)\} = p \in \mathfrak{N}^{\delta} .$$

<u>Proof</u>. We may assume that δ is implemented by a unitary w_π such that

$$(\pi(\varphi)\xi|\eta) = \langle w_\pi, \omega_{\xi,\eta} \otimes \varphi\rangle , \quad \varphi \in A(G) , \quad \xi,\eta \in \mathfrak{K} ;$$

$$\delta(a) = w_\pi^*(a \otimes 1)w_\pi \quad ;$$

$$w_\pi \in \pi(A(G))'' \bar{\otimes} \mathcal{R}(G) .$$

Let $Y = \{\delta_\varphi(x) : \varphi \in A(G)\}$. Then Y is a subspace of $\mathfrak{n}$ invariant under δ_ψ, $\psi \in A(G)$. Suppose $p\xi = 0$, $\xi \in \mathfrak{K}$. For any $\varphi = \omega_{f,g} \in A(G)$, $\eta \in \mathfrak{K}$ and $y \in Y$, we have

$$(y\pi(\varphi)\xi \mid \eta) = (\pi(\varphi)\xi \mid y^*\eta) = \langle w_\pi, \omega_{\xi,y^*\eta} \otimes \varphi\rangle$$

$$= \langle w_\pi, \omega_{\xi,\eta}\, y \otimes \varphi\rangle = \langle (y \otimes 1)w_\pi, \omega_{\xi,\eta} \otimes \varphi\rangle$$

$$= \langle w_\pi \delta(y), \omega_{\xi,\eta} \otimes \omega_{f,g}\rangle = (w_\pi \delta(y)(\xi \otimes f) \mid \eta \otimes g) .$$

For any $\zeta \in \mathfrak{K}$ and $h \in L^2(G)$, we have

$$(\delta(y)(\xi \otimes f) \mid \zeta \otimes h) = \langle \delta(y), \omega_{\xi,\zeta} \otimes \omega_{f,h}\rangle$$

$$= \langle \delta_\psi(y), \omega_{\xi,\zeta}\rangle \qquad \text{with } \psi = \omega_{f,h} ,$$

$$= (\delta_\psi(y)\xi \mid \zeta) = 0 ;$$

thus $\delta(y)(\xi \otimes f) = 0$; so that $(y\pi(\varphi)\xi \mid \eta) = 0$ for every $\eta \in \mathfrak{K}$; hence $y\pi(\varphi)\xi = 0$. Namely we have $p\pi(\varphi)\xi = 0$. Hence p and $\pi(\varphi)$ commute because $\pi(\varphi)$ leaves the $(1-p)\mathfrak{K}$ invariant. Hence $(p \otimes 1)$ and w_π commute, which means $\delta(p) = w_\pi^*(p \otimes 1)W_\pi = p \otimes 1$.
Q.E.D.

<u>Corollary</u> 2.7 If $\Gamma(\delta)$ is normal, then $\Gamma(\delta) = \Gamma(\tilde{\delta})$. In particular, $\Gamma(\delta) = G$ if and only if $\Gamma(\tilde{\delta}) = G$.

§3. Action of G on the cohomology space.

In this section we assume the proper infiniteness for $\mathfrak{n}$ and shall only give a sketch how to make an action on the cohomology space which relates to the Connes spectrum $\Gamma(\delta)$. For convenience we simply denote $Z_\delta(G,\mathfrak{n})$ by Z_δ.

Let $\mathfrak{D} = \mathfrak{n} \bar{\otimes} \mathfrak{L}(\ell^2(Z_\delta))$ and $\{e_{a,b} : a,b \in Z_\delta\}$ the canonical matrix units in $\mathfrak{L}(\ell^2(Z_\delta))$. We define a 1-cocycle u in $Z_{\bar{\delta}}(G,\mathfrak{D})$ $(\bar{\delta} = (\iota \otimes \sigma) \circ (\delta \otimes \iota))$ by

$$u = (\iota \otimes \sigma)(\sum_{a \in Z_\delta} a \otimes e_{aa}) . \tag{3.1}$$

Then, the map

$$\Xi = {}_u\bar{\delta} \;(= \mathrm{Ad}_u \circ (\iota \otimes \sigma) \circ (\delta \otimes \iota))$$

is also a co-action of G on $\mathfrak{D}$. We denote by ρ the center of $\mathfrak{D}^{\Xi}$.

For each $a \in Z_\delta(G,\mathfrak{n})$ we denote by $p_{\mathfrak{n}}(a)$ the support projection of $e_a \otimes e_{aa}$ in ρ.

Proposition 3.1. (i) The map $p_{\mathfrak{n}}$ transforms $Z_\delta(G,\mathfrak{n})$ onto all σ-finite projections in ρ.

(ii) $p_{\mathfrak{n}}(a) = p_{\mathfrak{n}}(\check{a})$.

(iii) $p_{\mathfrak{n}}(a) \le p_{\mathfrak{n}}(b)$ if and only if $\check{a} \prec \check{b}$.

Proof. (ii) That $p_{\mathfrak{n}}(a) \le p_{\mathfrak{n}}(\check{a})$ is clear. Since $\check{a} = \Sigma a_j$ with $a_j = (v_j \otimes 1)a\,\delta(v_j)^*$ by (1.6),

$$p_{\mathfrak{n}}(\check{a}) = \sum_b c_b(\check{a}) \otimes e_{bb} = \sum_b (\bigvee_j c_b(a_j)) \otimes e_{bb} \le p_{\mathfrak{n}}(a) .$$

(iii) That $p_{\mathfrak{n}}(a) \le p_{\mathfrak{n}}(b)$ if and only if $p_{\mathfrak{n}}(\check{a}) \le p_{\mathfrak{n}}(\check{b})$, if and only if $c_b(\check{a}) \le c_b(\check{b})$, if and only if $\check{a} \prec \check{b}$ by Proposition 1.5.

(i) Let $\{e_i : i \in I\}$ be a family of mutually orthogonal non-zero projections in ρ with $e_i \le p_{\mathfrak{n}}(a)$. For each $i \in I$, $e_i(e_a \otimes e_{aa}) \ne 0$. Since $\mathfrak{n}^a$ is σ-finite, I is countable.

Conversely, we shall show that each σ-finite projection e in ρ is of the form $p_{\mathfrak{n}}(a)$ for some $a \in Z_\delta(G,\mathfrak{n})$. Since the left support $s_\ell(u)$ of u defined by (3.1) is $\Sigma e_a \otimes e_{aa}$ and $e \le s_\ell(u)$, it follows that

$$e \leq \bigvee \{p_{\eta}(a) : a \in Z_{\delta}(G,\eta)\}.$$

Since e is σ-finite, there exists a sequence $\{a_j : j \in \mathbb{N}\}$ in $Z_{\delta}(G,\eta)$ such that

$$e \leq \bigvee \{p_{\eta}(a_j) : j \in \mathbb{N}\}.$$

Take a 1-cocycle $b \in Z_{\delta}(G,\eta)$ such that $a_j \prec b$ for all $j \in \mathbb{N}$. Then $e \leq p_{\eta}(b)$. Choose a 1-cocycle $a \in Z_{\delta}(G,\eta)$ such that

$$e(e_b \otimes e_{bb}) = c_b(a) \otimes e_{bb} .$$

Now, e and $p_{\eta}(a)$ are both projections in $\mathfrak{D}^{\Xi}$ dominated by the central support $p_{\eta}(b)$ of $e_b \otimes e_{bb}$ and

$$e(e_b \otimes e_{bb}) = p_{\eta}(a)(e_b \otimes e_{bb}) .$$

Consequently, $e = p_{\eta}(a)$. Q.E.D.

<u>Proposition</u> 3.2 If $b \in Z_{\delta}(G,\eta)$, there exists uniquely an isomorphism p_b of the center of η^b onto ρ_d with $d = p_{\eta}(b)$ such that $p_b(e) = p_{\eta}((e \otimes 1)b)$ for $e \in C_{\eta^b}$. In particular, $p_b(c_b(a)) = p_{\eta}(a)p_{\eta}(b)$.

If a is a 1-cocycle in $Z_{\delta}(G,\eta)$, then $(1 \otimes \rho(r))a$ is also a 1-cocycle in $Z_{\delta}(G,\eta)$.

<u>Definition</u> 3.3 A homomorphism $\mathfrak{z}$ of G into $\mathrm{Aut}(\rho)$ is defined by

$$\mathfrak{z}_t p_{\eta}(a) = p_{\eta}((1 \otimes \rho(t))a)$$

for all $a \in Z_{\delta}(G,\eta)$.

<u>Theorem</u> 3.4 The following three conditions are equivalent for $a \in Z_{\delta}(G,\eta)$:

(i) a is integrable.

(ii) $a \prec b$ for some (or all) dominant $b \in Z_{\delta}(G,\eta)$.

(iii) The map $t \to \mathfrak{z}_t p_{\eta}(a)$ is σ-strongly continuous.

<u>Proof</u>. (i) $\Longleftrightarrow$ (ii) : By Theorem 2.4.

(ii) $\Rightarrow$ (iii) : Since $p_{\mathfrak{n}}(a) = p_{\mathfrak{n}}(\check{a})$ by Proposition 3.1, we may assume that $\mathfrak{n}^a$ is properly infinite. Then $a \simeq (e \otimes 1)b$ for some dominant 1-cocycle $b \in Z_\delta(G,\mathfrak{n})$ and some projection e in the center of $\mathfrak{n}^b$. Since b is dominant, it is dual and hence there exists a unitary representation u of G in $\mathfrak{n}$ such that ${}_b\delta(u(t)) = u(t) \otimes \rho(t)$. Since $a \simeq (e \otimes 1)b$ and $(1 \otimes \rho(t))a \simeq (1 \otimes \rho(t))(e \otimes 1)b$, it follows that

$$p_{\mathfrak{n}}(a) = p_{\mathfrak{n}}((e \otimes 1)b)$$

$$p_{\mathfrak{n}}((1 \otimes \rho(t))a) = p_{\mathfrak{n}}((u(t)eu(t)^* \otimes 1)b) \ . \tag{3.2}$$

Because, since $c\,b^* \in Z_{{}_b\delta}(G,\mathfrak{n})$, we find that

$${}_b\delta(u(t))c\delta(u(t))^* \in Z_\delta(G,\mathfrak{n})$$

by direct computation. Since

$$(1 \otimes \rho(t))c = (u(t)^* \otimes 1)\ {}_b\delta(u(t))\, c\delta(u(t))^*\, \delta(u(t)),$$

we have $(1 \otimes \rho(t))c \simeq {}_b\delta(u(t))c\,\delta(u(t))^*$. Here we replace c by $(e \otimes 1)b$ and obtain

$${}_b\delta(u(t))(e \otimes 1)b\,\delta(u(t))^* = (u(t)eu(t)^* \otimes 1)b \ .$$

Therefore (3.2) is true.

On the other hand, we have an isomorphism p_b of the center of $\mathfrak{n}^b$ onto $\mathcal{P}_d$ $(d = p_{\mathfrak{n}}(b))$ such that $p_b(e) = p_{\mathfrak{n}}((e \otimes 1)b)$ by Proposition 3.2. Therefore the map

$$p_{\mathfrak{n}}(a) = p_b(e) \mapsto \mathfrak{F}_t p_{\mathfrak{n}}(a) = p_b(u(t)eu(t)^*)$$

is σ-strongly continuous.

(iii) $\Rightarrow$ (ii) : Let G_0 be a dense countable subgroup of G. Let $e = p_{\mathfrak{n}}(a)$ and $f = \bigvee\{\mathfrak{F}_t(e) : t \in G_0\}$. By assumption, each $\mathfrak{F}_t(e)$ is a σ-strong limit of $\mathfrak{F}_{t_n}(e)$ for some sequence $\{t_n : n \in \mathbb{N}\}$ in G_0. Therefore, $f = \bigvee\{\mathfrak{F}_t(e) : t \in G\}$ and it is σ-finite and $\mathfrak{F}$ invariant. By Proposition 3.1 we have $f = p_{\mathfrak{n}}(b)$ for some dominant $b \in Z_\delta(G,\mathfrak{n})$. Since $e \leq f \leq p_{\mathfrak{n}}(b)$,

$a \prec \check{a} \prec b$ by Proposition 3.1.

Q.E.D.

Corollary 3.5. If $\mathcal{M}^{\delta}$ is properly infinite, there exists a dual $a \in Z_{\delta}(G,\mathcal{M})$ such that $\{\mathcal{M}, {}_{a}\delta\} \cong \{\mathcal{R} \times_{\alpha} G, \hat{\alpha}\}$ and

$$\{\mathcal{P}_{d}, \mathfrak{F}\restriction d\} \cong \{C_{\mathcal{M}}, \alpha\},$$

where $d = p_{\mathcal{M}}(a)$ and $\mathfrak{F}\restriction d$ is the restriction of $\mathfrak{F}$ to $\mathcal{P}_{d}$. Thus the normalized Connes spectrum $\Gamma_{n}(\delta)$ is precisely the kernel of $\mathfrak{F}\restriction d$.

NOTES

The contents of this chapter are generalizations of the flow of weights on a factor of type III in [14].

CHAPTER VI

RELATIVE COMMUTANT OF CROSSED PRODUCTS.

Introduction. The relative commutant of the original algebra in the crossed product plays an important role in the study of the crossed product and/or the fixed point algebra. Section 1 is devoted to the analysis of the condition under which the relative commutant is contained in the original algebra. In the discrete case, the relative commutant property, (1.1), is equivalent to the freeness of the action of G. But, in the continuous case, the freeness of each individual group element does not yield (1.1) as in the example following Theorem 1.1. Theorem 1.1 gives a criteria for a stronger relative commutant property, i.e. a necessary and sufficient condition for the relative commutant of the center of the original algebra to be the original algebra itself. Generally speaking, the results in this area are very primitive. We need a greater effort to understand the relative commutant property to claim anything definite.

In §2, we treat the stability of actions following Connes-Takesaki, [14]. Again we know very little about it.

§1. Relative commutant theorem.

It is known that, if G is descrete, the following two conditions are equivalent:

(i) α_t is free on $\mathfrak{m}$ for each $t \neq e$ ($x\alpha_t(y) = yx$ for all y implies $x = 0$);

(ii) $\alpha(\mathfrak{m})' \cap (\mathfrak{m} \times_\alpha G) \subset \alpha(\mathfrak{m})$.

Indeed, $\sum_{t \in G} \alpha(x_t)(1 \otimes \rho(t)) \in \alpha(\mathfrak{m})' \cap (\mathfrak{m} \times_\alpha G)$, if and only if $x_t\alpha_t(y) = y x_t$ for all $y \in \mathfrak{m}$. However, the above equivalence does not hold for continuous group.

Our interest in this section is mainly concerned with

$$\alpha(\mathfrak{m})' \cap (\mathfrak{m} \times_\alpha G) = \alpha(\mathcal{C}_\mathfrak{m}). \tag{1.1}$$

Theorem 1.1 The following conditions for α are equivalent:

(i) $\alpha(\mathcal{C}_\mathfrak{m})' \cap (\mathfrak{m} \times_\alpha G) = \alpha(\mathfrak{m})$;

(ii) For any compact subset K of G, which does not contain the unit element $e \in G$, and any non-zero projection $p \in \mathcal{C}_\mathfrak{m}$ there exists a non-zero projection $q \leq p$ such that $q \in \mathcal{C}_\mathfrak{m}$ and

$$\alpha_t(q)q = 0 \quad \text{for every} \quad t \in K;$$

(iii) If $\{\Gamma,\mu\}$ is a standard G-measure space such that the canonical action of G on $L^\infty(\Gamma,\mu)$ is isomorphic to $\{\mathcal{C}_\mathfrak{m},G,\alpha\}$, then there exists a Borel null set N in Γ such that for every $\gamma \in \Gamma - N$, the stabilizer subgroup H_γ of γ reduces to $\{e\}$.

Proof. (iii) $\Rightarrow$ (ii): Let A be a globally α invariant σ-weekly dense separable C^*-subalgebra of $\mathcal{C}_\mathfrak{m}$ such that $\lim_{t\to e} \|\alpha_t(x) - x\| = 0$ for every $x \in A$. Let Γ be the spectrum of A. Then G acts on Γ continuously in such a way that $\alpha_t(a)(\gamma) = a(t^{-1}\gamma)$, $a \in A$, $t \in G$, $\gamma \in \Gamma$. Furthermore, let μ be the Radon measure on Γ corresponding to a faithful normal state on $\mathcal{C}_\mathfrak{m}$. It then follows that the measure μ is quasi invariant under the action of G and $\{\mathcal{C}_\mathfrak{m},G,\alpha\}$ can be identified with the action of G on $L^\infty(\Gamma,\mu)$. Set

$$\Gamma_K = \{\gamma \in \Gamma : t\gamma \neq \gamma \text{ for every } t \in K\}$$

By condition (iii), $\mu(\Gamma - \Gamma_K) = 0$ and Γ_K is open in Γ by the continuity of the action of G on Γ. For each $\gamma \in \Gamma_K$ and $t \in K$, there exists a compact neighborhood U_t of γ such that $tU_t \cap U_t = \emptyset$. Then we can choose a neighborhood $V(t)$ of t such that $V(t)U_t \cap U_t = \emptyset$. Then $\{V(t) : t \in K\}$ is an open covering of the compact set K, so that we can choose a finite subcovering $\{V(t_1), \dots, V(t_n)\}$ of K. Set $U(\gamma) = \bigcap_{i=1}^{n} U_{t_i}$. Then $U(\gamma)$ is a neighborhood of γ such that

$$KU(\gamma) \cap U(\gamma) = \emptyset . \tag{1.2}$$

Thus every point $\gamma \in \Gamma_K$ admits a neighborhood $U(\gamma)$ such that (1.2) holds. Condition (ii) now follows from this easily.

(ii) $\Rightarrow$ (i): Let $\rho = \alpha(C_{\mathfrak{m}})' \cap (\mathfrak{m} \times_\alpha G)$. We consider the restriction of the dual co-action $\hat{\alpha}$ to ρ. We claim that $\rho^{\hat{\alpha}}(K) = \{0\}$ for every compact subset K of G such that $e \notin K$. Let L be a compact subset of G such that K is contained in the interior of L and $e \notin L$. If $f \in A(G)$ such that $f(t) = 1$ for every $t \in K$ and $\mathrm{supp}(f) \subset L$, then $\hat{\alpha}_f(x) = x$ for every $x \in \rho^{\hat{\alpha}}(K)$. Let q be a projection such that $\alpha_t(q)q = 0$ for all $t \in L$. We then have

$$q\hat{\alpha}_f(x)q = 0 , \quad x \in \mathfrak{m} \times_\alpha G. \tag{1.3}$$

This follows from the fact that (1.3) holds for every x of the form $x = \int_G \alpha(x(t))(1 \otimes \rho(t))dt$. Hence, for every $x \in \rho^{\hat{\alpha}}(K)$, we have

$$xq = q\hat{\alpha}_f(x)q = 0 .$$

But condition (ii) says that $\vee\{q : \alpha_t(q)q = 0 \text{ for every } t \in L\} = 1$. Thus $\rho^{\hat{\alpha}}(K) = \{0\}$. Hence, $\rho = \rho^{\hat{\alpha}}(\{e\}) \subset \alpha(\mathfrak{m})$.

(i) $\Rightarrow$ (iii) : Set

$$\{\gamma \in \Gamma : H_\gamma \neq \{e\}\} = N .$$

Suppose $\mu(N) > 0$. Since N is the image of the Borel set

$$\{(t,\gamma) \in G \times \Gamma : t\gamma = \gamma, t \neq e\} \subset G \times \Gamma$$

under the projection to the second component Γ, N is analytic. By the von Neumann cross-section theorem, there exists a measurable G-valued function: $\gamma \in N \to h_\gamma \in G$ such that $h_\gamma \in H_\gamma - \{e\}$. We then define $h_\gamma = e$ for $\gamma \notin N$ and an operator u on

$\mathfrak{H} \bar{\otimes} L^2(G)$ as follows:

$$(u\xi)(\gamma,t) = \xi(\gamma, th_{t^{-1}\gamma}) , \quad \xi \in \mathfrak{H} \otimes L^2(G) ,$$

where we consider the central decomposition of $\mathfrak{H}$:

$$\mathfrak{H} = \int_{\Gamma}^{\oplus} \mathfrak{H}(\gamma) d\mu(\gamma) .$$

It is then easy to check that $u \in \alpha(\mathcal{C}_{\mathfrak{M}})' \cap (\mathfrak{M} \times_{\alpha} G)$ but $u \notin \alpha(\mathfrak{M})$.

Q.E.D.

An important consequence of the theorem is that the freeness of the action of each individual group element of G does not guarantee the relative commutant property, which shows a sharp contrast with the descrete case. A more precise example of this phenomena can be constructed as follows:

Consider $GL(n;\mathbf{R})$ action on $\mathbf{R}^n$ equipped with the Lebesgue measure m. Then consider any non-transitive ergodic group free action $\{\Gamma_1, \mu_1, G_1\}$. Set $G = G_1 \times GL(n;\mathbf{R})$, $\Gamma = \Gamma_1 \times \mathbf{R}^n$, and $\mu = \mu_1 \times m$. We then get a non-transitive ergodic transformation group such that each single group element acts freely on $\{\Gamma,\mu\}$. But it does not have the relative commutant property.

<u>Proposition</u> 1.2. Let $\mathfrak{a}$ be an abelian von Neumann algebra equipped with an action α of G. If $\mathrm{Aut}_{\alpha}(\mathfrak{a}) = \{\sigma \in \mathrm{Aut}(\mathfrak{a}) : \sigma\alpha_t = \alpha_t\sigma, \ t \in G\}$ is ergodic on $\mathfrak{a}$ and if α is faithful, then the action α satisfies the conditions in Theorem 1.1 for $\{\mathcal{C}_{\mathfrak{M}}, \alpha\}$.

<u>Proof</u>. Let $\{\Gamma,\mu\}$ be the G-measure space considered in the proof (iii) $\Rightarrow$ (ii) of Theorem 1.1. We note that $H = \mathrm{Aut}_{\alpha}(\mathfrak{a})$ acts on $\{\Gamma,\mu\}$ also and commutes with the action of G. Set

$$N = \{\gamma \in \Gamma : H_{\gamma} \neq \{e\}\} .$$

It then follows that N is invariant under H; hence either $\mu(N) = 0$ or $\mu(\Gamma - N) = 0$. For each compact set K in G with $e \notin K$, set

$$\Gamma_K = \{\gamma \in \Gamma : K \cap H_\gamma = \emptyset\}.$$

It follows that Γ_K is open and H-invariant. Hence either Γ_K is dense in Γ and $\mu(\Gamma_K) = 1$ or $\Gamma_K = \emptyset$. If $s \neq e$, then there exists $\gamma \in \Gamma$ such that $s\gamma \neq \gamma$, so that there exists a compact neighborhood $V(s)$ of s such that $\gamma \notin V(s)\gamma$, which means that $\gamma \in \Gamma_{V(s)}$. Hence $\Gamma_{V(s)}$ is dense in Γ. Choosing a finite covering $\{V(s_i) : i = 1,2,\dots,n\}$ of K, we have

$$\Gamma_K \supset \Gamma_{V(s_1) \cup \dots \cup V(s_n)} = \bigcap_{i=1}^{n} \Gamma_{V(s_i)} .$$

By the Baire property of Γ, we have $\dot{\Gamma}_K \neq \emptyset$, so that $\mu(\Gamma_K) = 1$ for every compact subset K of G with $e \notin K$. If $\{K_n\}$ is an increasing sequence of compact subsets of G such that $G - \{e\} = \bigcup_{n=1}^{\infty} K_n$, then we have $\Gamma - N = \bigcap_{n=1}^{\infty} \Gamma_{K_n}$. Since $\mu(\Gamma - N) = \lim \mu(\Gamma_{K_n}) = 1$, N must be null. Q.E.D.

<u>Corollary</u> 1.3. If G is abelian, then the ergodicity of α on $\mathcal{C}_{\mathfrak{m}}$ together with the faithfulness of the restriction of α to $\mathcal{C}_{\mathfrak{m}}$ yields the conditions in Theorem 1.1.

Denote by $Z_\alpha(G,\mathcal{U}(\mathcal{C}_{\mathfrak{m}}))$ the set of all unitary 1-cocycles in $Z_\alpha(G,\mathcal{C}_{\mathfrak{m}})$. For each $a \in Z_\alpha(G,\mathcal{U}(\mathcal{C}_{\mathfrak{m}}))$ we set $\beta_a = \mathrm{Ad}_{a*} \upharpoonright \mathfrak{m} \times_\alpha G$. Then, by direct computation,

$$\beta_a(\alpha(x)) = \alpha(x) \ , \ \beta_a(1 \otimes \rho(r)) = \alpha(a(r))(1 \otimes \rho(r)) .$$

<u>Theorem</u> 1.4. If α satisfies (1.1), then

(i) The above map : $a \to \beta_a$ is a bijective isomorphism of $Z_\alpha(G,\mathcal{U}(\mathcal{C}_{\mathfrak{m}}))$ onto $\mathrm{Aut}(\mathfrak{m} \times_\alpha G/\alpha(\mathfrak{m}))$.

(ii) $a \in B_\alpha(G,\mathcal{U}(\mathcal{C}_{\mathfrak{m}}))$(i.e. $a \simeq 1$) if and only if $\beta_a \in \mathrm{Int}(\mathfrak{m} \times_\alpha G)$.

<u>Proof</u>. (i) Suppose that $\beta \in \mathrm{Aut}(\mathfrak{m} \times_\alpha G/\alpha(\mathfrak{m}))$. We set

$$b(r) = \beta(1 \otimes \rho(r))(1 \otimes \rho(r))^* .$$

Since $\mathrm{Ad}_{b(t)}(\alpha(x)) = \alpha(x)$ for $x \in \mathfrak{m}$, $b(t)$ belongs to $\alpha(\mathfrak{m})' \cap (\mathfrak{m} \times_\alpha G)$. Since α satisfies (1.1), we have $b(t) \in \alpha(\mathcal{C}_{\mathfrak{m}})$ and so $b(t) = \alpha(a(t))$ for some $a(t) \in \mathcal{C}_{\mathfrak{m}}$. Since $t \to b(t)$ is strongly continuous and $b(st) = b(s)(\iota \otimes \rho_s)(b(t))$, it follows that a is a unitary 1-cocycle in $Z_\alpha(G,\mathcal{C}_{\mathfrak{m}})$ with $\beta = \beta_a$.

(ii) That $\beta_a \in \mathrm{Int}(\mathfrak{m} \times_\alpha G)$, i.e. $\beta_a = \mathrm{Ad}_u$ for some unitary u in $\mathfrak{m} \times_\alpha G$ is equivalent to $u \in \alpha(\mathfrak{m})' \cap (\mathfrak{m} \times_\alpha G)$ and $u(1 \otimes \rho(r))u^* = \alpha(a(r))(\iota \otimes \rho(r))$ for $r \in G$. Since α satisfies (1.1), this is equivalent to $u = \alpha(v)$ for some $v \in \mathcal{C}_{\mathfrak{m}}$ and $v\,\alpha_r(v^*) = a(r)$. Q.E.D.

Proposition 1.5. (i) If α is integrable and satisfies (1.1), then

(1.4) $$(\mathfrak{M}^{\alpha})' \cap \mathfrak{M} = \mathbf{C}_{\mathfrak{M}} .$$

(ii) If G is abelian and α is dual, then (1.4) implies (1.1).

Proof. (i) Let $\varphi = \omega \circ \mathcal{E}_{\alpha}$ for some faithful normal state ω on $\mathfrak{M}$. As α is integrable, φ is a faithful, semi-finite, normal weight on $\mathfrak{M}$ such that $\varphi \circ \alpha_t = \varphi$. We may assume that $\mathfrak{M}$ acts standardly on the L^2 completion of $\mathfrak{n}_{\varphi}$. Let J be the modular unitary involution and u a unitary in $\mathcal{L}(\mathfrak{H}) \bar{\otimes} L^{\infty}(G)$ defined by $u(t)\eta_{\varphi}(x) = \eta_{\varphi}(\alpha_t(x))$ for $x \in \mathfrak{n}_{\varphi}$. Since α is integrable, the map: $t \to u(t)\xi, \xi \in \mathfrak{n}_{\varphi}$ is square integrable. Let U be an isometry of $\mathfrak{H}$ into $\mathfrak{H} \otimes L^2(G)$ defined by

$$(U\xi)(t) = u(t)\xi, \qquad \xi \in \mathfrak{H} .$$

Then we have, for any $x \in \mathcal{L}(\mathfrak{H})$ and $t \in G$,

(1.5) $$Ux = u(x \otimes 1)u^* U , \quad Uu(t) = (1 \otimes \rho(t))U.$$

Since $J((\mathfrak{M}^{\alpha})' \cap \mathfrak{M})J = (\mathfrak{M} \vee u(G)'') \cap \mathfrak{M}'$ and

$$(\alpha(\mathfrak{M}) \vee (\mathbf{C} \otimes \mathcal{R}(G)'') \cap u(\mathfrak{M}' \otimes \mathbf{C})u^*$$

$$\subset (\mathfrak{M} \times_{\alpha} G) \cap \alpha(\mathfrak{M})' = \alpha(\mathbf{C}_{\mathfrak{M}})$$

by assumption (1.1), if $y \in (\mathfrak{M}^{\alpha})' \cap \mathfrak{M}$, then $UJyJ \in \alpha(\mathbf{C}_{\mathfrak{M}})U = U\mathbf{C}_{\mathfrak{M}}$ by (1.5). Since U is isometry, $JyJ \in \mathbf{C}_{\mathfrak{M}}$ and hence $y \in \mathbf{C}_{\mathfrak{M}}$.

(ii) We may assume that $\mathfrak{M}$ is standard. Let u be a unitary representation of G on $\mathfrak{H}$ implementing canonically α, [1,24]. We define a map U_f of $\mathfrak{H}$ into $\mathfrak{H} \otimes L^2(G)$ by

$$U_f\xi = u(\xi \otimes f), \qquad \xi \in \mathfrak{H}, \quad f \in L^2(G).$$

Then we have

(1.6) $$u(x \otimes 1)u^* U_f = U_f x , \qquad x \in \mathcal{L}(\mathfrak{H}).$$

First, we notice that, if $y \in \alpha(\mathfrak{M})' \cap (\mathfrak{M} \times_{\alpha} G)$, then

(1.7) $$JU_g^* y U_f J \in (\mathfrak{M}^{\alpha})' \cap \mathfrak{M} \; (= \mathbf{C}_{\mathfrak{M}}) ,$$

where J is the modular unitary involution for $\mathfrak{M}$. Indeed, since $y \in \alpha(\mathfrak{M})'$, (1.6) implies that $U_g^* y U_f \in \mathfrak{m}'$. Since $y \in \mathfrak{m} \times_\alpha G$, (1.6) implies $U_g^* y U_f \in \mathfrak{m} \vee u(G)''$, because, if $x \in \mathfrak{m}' \cap u(G)'$, then $U_f x = u(x \otimes 1)u^* U_f = (x \otimes 1)U_f$ and hence $[U_g^* y U_f, x] = 0$.

We now show (1.1). By Theorem II. 3.2. (a) we have only to show that

$$y \in \alpha(\mathfrak{m})' \cap (\mathfrak{m} \times_\alpha G) \Rightarrow y \in (\mathbb{C} \otimes L^\infty(G))'.$$

By assumption, G is abelian and α is dual. Therefore for each $p \in \hat{G}$ there exists a unitary $v_p \in \mathfrak{m}$ such that $\alpha_t(v_p) = \overline{\langle t, p\rangle}\, v_p$. Put $v'_p = J v_p J \in \mathfrak{m}'$. Since $Ju(t)J = u(t)$, we have $u(t)v'_p u(t)^* = \langle t,p\rangle v'_p$ and

$$(1.8) \qquad u(v'_p \xi \otimes f) = (v'_p \otimes 1)u(\xi \otimes pf).$$

If $y \in \alpha(\mathfrak{m})' \cap (\mathfrak{m} \times_\alpha G)$, then

$$(y(1 \otimes p)u(\xi \otimes f) | u(\eta \otimes g)) = (yu(\xi \otimes pf) | u(\eta \otimes g))$$

$$= (y(v'_p \otimes 1)^* u(v'_p \otimes 1)(\xi \otimes f) | u(\eta \otimes g)) \qquad \text{by}(1.8),$$

$$= (yu(v'_p\xi \otimes f) | (v'_p \otimes 1)u(\eta \otimes g))$$

$$= (yu(v'_p\xi \otimes f) | u(v'_p\eta \otimes \bar{p}g)) \qquad \text{by}(1.8)$$

$$= (yU_f v'_p \xi | U_{\bar{p}g} v'_p \eta)$$

$$= (yU_f \xi | U_{\bar{p}g}\eta) \qquad \text{by}(1.7)$$

$$= ((1 \otimes p)yu(\xi \otimes f) | u(\eta \otimes g)).$$

Thus $[y, 1 \otimes p] = 0$ for all $p \in \hat{G}$ and hence $y \in (\mathbb{C} \otimes L^\infty(G))'$. Q.E.D.

Proposition 1.6. (i) If α is integrable and satisfies the relative commutant property:

$$(1.9) \qquad \alpha(\mathfrak{m})' \cap (\mathfrak{m} \times_\alpha G) \subset \mathfrak{C}_{\mathfrak{m} \times_\alpha G},$$

then

$$(1.10) \qquad (\mathfrak{m}^\alpha)' \cap \mathfrak{m} = \mathbb{C}_{\mathfrak{m}^\alpha}.$$

(ii) If G is abelian, then (1.10) implies (1.9).

Proof. (i) We use the same device and notations as in the proof of Proposition 1.5.i. If $y \in (\mathfrak{m}^\alpha)' \cap \mathfrak{m}$, then $UJyJ \in \mathfrak{C}_{\mathfrak{m} \times_\alpha G} U$ by (1.5). Let C be the complex conjugation operator on $L^2(G)$. Then

$$UJu(t)J = (J \otimes C)(1 \otimes \rho(t))(J \otimes C)U = (1 \otimes \rho(t))U.$$

Therefore JyJ commutes with $Ju(t)J$, and hence $y \in \mathfrak{M}^{\alpha}$.

(ii) We use the same notations as in the proof of Proposition 1.5. (ii). First we notice that

$$(1.11) \qquad (1 \otimes \rho(r))\, U_f = U_{\rho(r)f}u(r)$$

$$(1.12) \qquad (u(r) \otimes \lambda(r))\, U_f = U_{\lambda(r)f}\,.$$

If $y \in \alpha(\mathfrak{M})' \cap (\mathfrak{M} \times_{\alpha} G)$, then $JU_g^* yU_fJ \in \mathfrak{M}^{\alpha}$ by (1.7). Since $Ju(t)J = u(t)$, this implies that

$$(1.13) \qquad U_g^* \, y \, U_f \in \mathfrak{M}^{\alpha}.$$

Since $y \in \alpha(\mathfrak{M})'$, it suffices to show that $y \in (\mathbb{C} \otimes \mathcal{R}(G))'$. By (1.11) and (1.13) we have

$$U_g^* \, y(1 \otimes \rho(r))U_f = U_{\rho(r)*g}^*(1 \otimes \rho(r))y\,U_{\rho(r)f}.$$

Since $y \in \mathfrak{M} \times_{\alpha} G$ and G is abelian, (1.11) implies

$$U_{\rho(r)*g}^*(1 \otimes \rho(r))y\,U_{\rho(r)f} = U_g^*(1 \otimes \rho(r))y\,U_f\,.$$

Therefore $y \in (\mathbb{C} \otimes \mathcal{R}(G))'$.

Q.E.D.

It should be noted that (1.10) is true for any integrable modular automorphisms, [14].

NOTES

The equivalence (i) $\Longleftrightarrow$ (ii) in Theorem 1.1 was obtained independently in [14,57]. Theorem 1.4 is taken from [14]. Propositions 1.5 and 1.6 are due to Paschke, [54].

§2. Stability

In this section we shall discuss the stability of actions or co-actions under 1-cocycle perturbations.

Definition 2.1. An action α (resp. co-action δ) is said to be stable, if all unitary 1-cocycles are equivalent, that is, for each unitary 1-cocycle a there exists a unitary $c \in \mathfrak{M}$ (resp. $\mathfrak{N}$) such that $a = (c^* \otimes 1)\,\alpha(c)$ (resp. $(c^* \otimes 1)\delta(c)$).

Lemma 2.2. If α satisfies (1.1), then (1.1) holds for both of the following:

(i) ${}_a\alpha$ for each unitary $a \in Z_\alpha(G,\mathfrak{M})$;

(ii) $\bar{\alpha}$ on $\mathfrak{M} \bar{\otimes} F_n$ ($n \in \mathbb{N}$ or ∞) given by $\bar{\alpha}_t = \alpha_t \otimes \iota$.

Proof. (i) We may assume that $\mathfrak{M}$ is standard and $\alpha(x) = u(x \otimes 1)u^*$ for some unitary $u \in \mathfrak{L}(\mathfrak{H}) \bar{\otimes} L^\infty(G)$ with $(u \otimes 1)(\iota \otimes \sigma)(u \otimes 1) = \alpha_G(u)$. Since

$$({}_a\alpha(\mathfrak{M}))' = (\mathrm{Ad}_{au}(\mathfrak{M} \otimes \mathbf{C}))' = a\,\alpha(\mathfrak{M})'a^*$$

$${}_a\alpha(\mathfrak{M}) \vee (\mathbf{C} \otimes \mathfrak{R}(G)) = a(\mathfrak{M} \times_\alpha G)a^*,$$

it follows that ${}_a\alpha$ satisfies (1.1) on $\mathfrak{M}$.

(ii) Since

$$(\alpha(\mathfrak{M}) \bar{\otimes} F_n)' \cap ((\alpha(\mathfrak{M}) \bar{\otimes} F_n) \vee (\mathbf{C} \otimes \mathfrak{R}(G) \otimes \mathbf{C})) \subset \alpha(\mathfrak{M}) \bar{\otimes} F_n$$

$$(\iota \otimes \sigma)((\alpha(\mathfrak{M}) \bar{\otimes} F_n)' = \bar{\alpha}(\mathfrak{M} \bar{\otimes} F_n)'$$

it follows that $\bar{\alpha}$ satisfies (1.1) on $\mathfrak{M} \bar{\otimes} F_n$.

Q.E.D.

Proposition 2.3 If α satisfies (1.1) and $\mathfrak{M}^a$ is properly infinite for all unitaries $a \in Z_\alpha(G, \mathfrak{M})$, then α is stable.

Proof. Take a unitary $a \in Z_\alpha(G,\mathfrak{M})$. We set $\bar{\mathfrak{M}} = \mathfrak{M} \otimes F_2$, $c(t) = 1 \otimes e_{11} + a(t) \otimes e_{22}$ and $\bar{\alpha}_t = \alpha_t \otimes \iota$. Then, by Lemma 2.2, ${}_c\bar{\alpha}$ satisfies (1.1) on $\mathfrak{M} \otimes F_2$. Therefore, Proposition 1.3 implies that $(\bar{\mathfrak{M}}^c)' \cap \bar{\mathfrak{M}} = \mathbf{C}_{\bar{\mathfrak{M}}}$, which implies that the center of $\bar{\mathfrak{M}}^c$ coincides with $\mathbf{C}_{\bar{\mathfrak{M}}} \cap \bar{\mathfrak{M}}^c$. Let e_j be the central supports of $1 \otimes e_{jj}$ in $\bar{\mathfrak{M}}^c$ for $j = 1,2$. Then $e_j \in \mathbf{C}_{\bar{\mathfrak{M}}} \cap \bar{\mathfrak{M}}^c$. Since $1 \otimes e_{11} \sim 1 \otimes e_{22}$ in $\bar{\mathfrak{M}}$, the fact $e_j \in \mathbf{C}_{\bar{\mathfrak{M}}}$ implies that $1 \otimes e_{11} + 1 \otimes e_{22} \leq e_j$. Thus $e_1 = 1 \otimes e_{11} + 1 \otimes e_{22} = e_2$, namely, $1 \otimes e_{11}$ and $1 \otimes e_{22}$ have the same central support in $\bar{\mathfrak{M}}^c$. As we assumed that $\mathfrak{M}^a$ is properly infinite for every unitary $a \in Z_\alpha(G,\mathfrak{M})$, $1 \otimes e_{11}$ and $1 \otimes e_{22}$ are properly infinite in $\bar{\mathfrak{M}}^c$.

Consequently, $1 \otimes e_{11} \sim 1 \otimes e_{22}$ in $\bar{\mathfrak{m}}^c$, namely, $a \simeq 1$.

Corollary 2.4. If G is finite and α_t is free for each $t \neq e$, then α is stable.

Proof. The case where $\mathfrak{m}$ is properly infinite. Since G is finite, $\mathfrak{m}^a$ is properly infinite for all unitary $a \in Z_\alpha(G,\mathfrak{m})$. Since α_t is free for all $t \neq e$, α satisfies (1.1). Therefore, α is stable by Proposition 2.3.

The case where $\mathfrak{m}$ is finite. Let $\bar{\mathfrak{m}} = \mathfrak{m} \bar{\otimes} F_\infty$ and $\bar{\alpha}_t = \alpha_t \otimes \iota$. Then $\bar{\alpha}$ satisfies (1.1) by Lemma 2.2, and hence the center of $\bar{\mathfrak{m}}^{\bar{\alpha}}$ is the $C_{\bar{\mathfrak{m}}} \cap \bar{\mathfrak{m}}^{\bar{\alpha}}$ by Proposition 1.3, which implies that $C_{\mathfrak{m}^\alpha} = C_{\mathfrak{m}} \cap \mathfrak{m}^\alpha$. Let $\mathcal{C}$ be the center valued trace on $\mathfrak{m}$. Then the restriction of $\mathcal{C}$ to $\mathfrak{m}^\alpha$ is also the center valued trace on $\mathfrak{m}^\alpha$. Therefore, for any projections e and f in $\mathfrak{m}^\alpha$,

$$e \sim f \text{ in } \mathfrak{m} \Longleftrightarrow e \sim f \text{ in } \mathfrak{m}^\alpha .$$

Thus $e_a \otimes e_{11} \sim e_b \otimes e_{22}$ for all unitaries a, b in $Z_\alpha(G,\mathfrak{m})$. Consequently, $a \simeq b$.

Q.E.D.

NOTES

The stability of an action played an important role in the structure analysis of a factor of type III, [14]. However, we have to admit that the general study of the stability of an action is in a very primitive stage. The authors feel that one should study also the approximate stability of actions. For example, the modular automorphism action of $\mathbb{R}$ on a factor of type III_1 is approximately stable by Connes-Størmer [13]. The materials presented here are taken from [14].

CHAPTER VII

APPLICATIONS TO GALOIS THEORY

Introduction Galois theory here means the theory concerning intermediate von Neumann sublagebras between the entire von Neuamnn algebra and the fixed point subalgebra under a prescribed action (resp. co-action) of G. The theory is intimately related to the analysis of the algebra of observables in the axiomatic approach to the quantum field theory of Doplicher-Haag-Roberts, [17]. It should, however, be pointed out that since our theory is concerned with von Neumann algebras, but not C*-algebras, we have still a long way to go in order to build a theory applicable directly to physics. Nevertheless, the study in the von Neumann algebra context would provide a way to go further.

In §1, we shall study the crossed product by a closed subgroup H or the right homogeneous space $H\backslash G$ according to an action or a co-action. Theorems 1.1 and 1.2 tell how one can recover a given closed subgroup H of G by the crossed product subalgebras, i.e. H is determined by $\mathfrak{M} \times_\alpha H$ (resp. $\mathfrak{N} \times_\delta (H\backslash G)$).

Section 2 is devoted to the association of closed subgroups H to intermediate von Neumann subalgebras in the crossed product under a natural regularity condition.

In §3, under regularity conditions we show Galois type correspondences between closed subgroups and intermediate subalgebras for semi-dual actions and co-actions. The regularity conditions are fullfilled when one has a sufficiently large automorphism group commuting with a given action of a compact group. Hence, if the situation arises from physics, and the action is given by the gauge group, then the space translation will provide the regularity condition. Making use of all the duality mechanisms, we shall show that an automorphism σ leaving $\mathfrak{M}^\alpha$ pointwise fixed must be in $\alpha(G)$ for a compact G if σ commutes with a large subgroup $\mathfrak{S}$ commuting with $\alpha(G)$, Theorem 3.8.

The last section is devoted to the full group analysis of Dye-Choda, which also provides a correspondence between certain subgroups, called full groups, of the automorphism group of the crossed product and intermediate subalgebras.

§1. Subgroups and crossed products

Let H be a closed subgroup of G. We shall give a characterization of $\mathfrak{m} \times_\alpha H$ and $\mathfrak{n} \times_\delta (H\backslash G)$ in Theorems 1.1 and 1.2. Then H will turn out to be the smallest closed subgroup K with $\hat{\alpha}(\mathfrak{m} \times_\alpha H) \subset \mathfrak{n} \bar{\otimes} \rho(K)''$ or the set of $t \in G$ with $\hat{\delta}_t = \iota$ on $\mathfrak{n} \times_\delta (H\backslash G)$. This indicates the one side of Galois type correspondences:

$$
\begin{array}{ccc}
G & & \mathfrak{m} \times_\alpha G \\
| & & | \\
H & \rightleftarrows & \mathfrak{m} \times_\alpha H \\
| & & | \\
\{e\} & & \alpha(\mathfrak{m}) ,
\end{array}
\qquad\qquad
\begin{array}{ccc}
\{e\} & & \mathfrak{n} \times_\delta G \\
| & & | \\
H & \rightleftarrows & \mathfrak{n} \times_\delta (H\backslash G) \\
| & & | \\
G & & \delta(\mathfrak{n})
\end{array} .
$$

We denote by ρ^H or λ^H the right or the left regular representation of H on $L^2(H)$.

Theorem 1.1. Let $\mathfrak{n} = \mathfrak{m} \times_\alpha G$. Then

(i) $\mathfrak{m} \times_\alpha H = \{y \in \mathfrak{n} : \hat{\alpha}(y) \in \mathfrak{n} \bar{\otimes} \rho(H)''\} = \mathfrak{n} \cap (\mathbf{C} \otimes \mathcal{L}^\infty(G/H))'$.

(ii) H is the smallest closed subgroup K satisfying $\hat{\alpha}(\mathfrak{m} \times_\alpha H) \subset \mathfrak{n} \bar{\otimes} \rho(K)''$.

Proof. (i) By (I.2.8) and (I.2.9) it is clear that $\hat{\alpha}(\mathfrak{m} \times_\alpha H)$ is contained in $\mathfrak{n} \bar{\otimes} \rho(H)''$.

Next we shall show that $\hat{\alpha}(y) \in \mathfrak{n} \bar{\otimes} \rho(H)''$ implies $y \in (\mathbf{C} \otimes \mathcal{L}^\infty(G/H))'$. For this it suffices to show that

$$
(1.1) \qquad \mathcal{L}^\infty(G/H) \otimes \mathbf{C} \subset (\mathrm{Ad}_{W_G}(\mathbf{C} \otimes \rho(H)')) \vee (\mathbf{C} \otimes \mathcal{L}(L^2(G))) .
$$

Indeed, $y \otimes 1$ commutes with $\mathrm{Ad}_{1 \otimes W_G}(\mathfrak{n}' \bar{\otimes} \rho(H)')$ and $\mathbf{C} \otimes \mathbf{C} \otimes \mathcal{L}(L^2(G))$ and hence y must commute with $\mathbf{C} \otimes \mathcal{L}^\infty(G/H)$ by (1.1). Now, if $f \in \mathcal{L}^\infty(G/H) \cap C(G)$, then $\alpha_G^!(f) \in \mathcal{L}^\infty(G/H) \bar{\otimes} L^\infty(G)$, where $(\alpha_G^! f)(s,t) = f(ts)$. For any $g,h \in K(G)$ with $\|\Delta g\|_1 = 1$ we set

$$
F_{g,h} = \left(\int \alpha_G^!(f_{r^{-1}})(1 \otimes g_{r^{-1}}) dr \right) (1 \otimes h) .
$$

Since $f \in \rho(H)'$ and $\alpha_G^!(f) = \mathrm{Ad}_{W_G}(1 \otimes f)$, $F_{g,h}$ belongs to the right hand side of (1.1). If $\xi \in L^2(G \times G)$, then

$$
(F_{g,h}\xi)(s,t) = \left(\int f(r^{-1}s) g(r^{-1}) dr \right) \Delta(t)^{-1} h(t) \xi(s,t) .
$$

Since $r \mapsto f(r^{-1}s)$ is continuous and bounded, if $g(r^{-1})dr$ converges to the Dirac measure at the unit, then $F_{g,h}$ converges weakly to $f \otimes (\Delta^{-1}h)$. Therefore $f \otimes (\Delta^{-1}h)$ belongs to the right hand side of (1.1). Making $\Delta^{-1}h$ converge weakly to 1, we have the inclusion (1.1) for $\mathcal{L}^\infty(G/H) \cap C(G)$ in place of $\mathcal{L}^\infty(G/H)$. As $\mathcal{L}^\infty(G/H) \cap C(G)$ is weakly dense in $\mathcal{L}^\infty(G/H)$, we have (1.1).

Finally we shall show that $\mathfrak{M} \times_\alpha H = \mathfrak{n} \cap (\mathbf{C} \otimes \mathcal{L}^\infty(G/H))'$. Suppose that $y \in \mathfrak{n} \cap (\mathbf{C} \otimes \mathcal{L}^\infty(G/H))'$. Here we may assume that $\mathfrak{M}$ is standard. Let $J_{\mathfrak{M}}$ and $\tilde{J}$ be the modular unitary involution of $\mathfrak{M}$ and $\mathfrak{M} \times_\alpha G$, respectively. It then follows from Proposition III.1.7 that

$$\tilde{J} = (J_{\mathfrak{M}} \otimes J)u^* = u(J_{\mathfrak{M}} \otimes J) ,$$

where u is the canonical unitary implementation of α. Then we have

$$\tilde{J}(\mathbf{C} \otimes \mathcal{L}^\infty(G/H))'\tilde{J} = (\mathbf{C} \otimes \mathcal{L}^\infty(H\backslash G))'$$

and hence $\tilde{J}y\tilde{J} \in \mathfrak{n}' \cap (\mathbf{C} \otimes \mathcal{L}^\infty(H\backslash G))'$. Here we apply the Blattner-Mackey's theorem for induced representations, [65]. There exists a natural isomorphism π of $(\mathfrak{M} \times_{\alpha^H} H)'$ onto $\mathfrak{n}' \cap (\mathbf{C} \otimes \mathcal{L}^\infty(H\backslash G))'$ such that

$$\pi(x' \otimes 1_H) = x' \otimes 1_G , \qquad x' \in \mathfrak{M}' ,$$

$$\pi(u(t) \otimes \lambda^H(t)) = u(t) \otimes \lambda(t) , \quad t \in H ,$$

where α^H is the restriction of α to H. Therefore, $\tilde{J}y\tilde{J}$ belongs to $\pi((\mathfrak{M} \times_{\alpha^H} H)')$, which is generated by $\mathfrak{M}' \otimes \mathbf{C}$ and $u(t) \otimes \lambda(t)$, $t \in H$. Since

$$\tilde{J}(x' \otimes 1)\tilde{J} = \alpha(J_{\mathfrak{M}}x'J_{\mathfrak{M}})$$

$$\tilde{J}(u(t) \otimes \lambda(t))\tilde{J} = 1 \otimes \rho(t) ,$$

it follows that y belongs to $\mathfrak{M} \times_\alpha H$.

(ii) It is clear from the first equality in (i). Q.E.D.

Theorem 1.2. Let $\mathfrak{M} = \mathfrak{n} \times_\delta G$. Then

(i) $\mathfrak{n} \times_\delta (H\backslash G) = \{x \in \mathfrak{M} : \hat{\delta}_t(x) = x, \ t \in H\} = \mathfrak{M} \cap (\mathbf{C} \otimes \lambda(H))'$.

(ii) $H = \{t \in G : \hat{\delta}_t(x) = x, \ x \in \mathfrak{n} \times_\delta (H\backslash G)\}$.

Proof. (i) It suffices to consider the commutant of (II.3.8).

(ii) Let K be the set of $t \in G$ such that $\hat{\delta}_t(x) = x$ for $x \in \mathfrak{n} \times_\delta (H\backslash G)$.

Then K is a closed subgroup of G containing H. Since $\mathfrak{n} \times_\delta (H\backslash G) \subset \mathfrak{n} \times_\delta (K\backslash G)$ by (i), we have $H = K$. Q.E.D.

NOTES

A Galois theory of von Neumann algebras are initiated by Nakamura and Takeda [51] and Suzuki [61] for discrete groups.

Theorems 1.1 and 1.2 are obtained by Takesaki [69] for abelian locally compact grups and Nakagami [47] for non abelian groups.

§2. Subalgebras in crossed products

Let $\mathcal{P}$ be a von Neumann subalgebra of a crossed product $\mathfrak{M} \times_\alpha G$ (resp. $\mathfrak{N} \times_\delta G$) containing $\alpha(\mathfrak{M})$ (resp. $\delta(\mathfrak{N})$). We shall give a condition for $\mathcal{P}$ to be of the form $\mathfrak{M} \times_\alpha H$ or $\mathfrak{N} \times_\delta (H\backslash G)$ for some closed subgroup H of G. In this case H is determined by $\mathcal{P}$ and $\mathcal{P}$ is determined by this H. Therefore, we have the other side of Galois type correspondences:

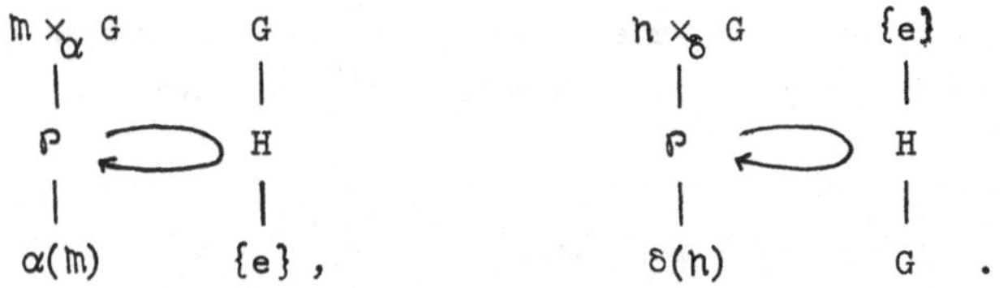

For the sake of a technical reason we shall assume that $\mathfrak{M}$ and $\mathfrak{N} \times_\delta G$ are factors in Theorems 2.1 and 2.2. In Theorem 2.3 we replace this assumption by other conditions.

<u>Theorem</u> 2.1. Let $\mathcal{P}$ be a von Neumann subalgebra of $\mathfrak{M} \times_\alpha G$ containing $\alpha(\mathfrak{M})$. If $\mathfrak{M}$ is a factor, then the following two conditions are equivalent:

(i) $\mathcal{P}$ is globally $\hat{\alpha}$ invariant.

(ii) $\mathcal{P} = \mathfrak{M} \times_\alpha H$ for some closed subgroup H of G.

<u>Proof</u>. (ii) $\Rightarrow$ (i): It is clear.

(i) $\Rightarrow$ (ii): Let $\mathfrak{N} = \mathfrak{M} \times_\alpha G$ and H the smallest closed subgroup of G satisfying $\hat{\alpha}(\mathcal{P}) \subset \mathfrak{N} \bar{\otimes} \rho(H)''$. Since $\hat{\alpha}(y) \in \mathfrak{N} \bar{\otimes} \rho(H)''$ is equivalent to $\mathrm{sp}_{\hat{\alpha}}(y) \subset H$, H coincides with the closed subgroup generated by $\mathrm{sp}_{\hat{\alpha}}(y)$ for all $y \in \mathcal{P}$. Therefore $\mathcal{P}$ is contained in the set of all $y \in \mathfrak{N}$ with $\hat{\alpha}(y) \in \mathfrak{N} \bar{\otimes} \rho(H)''$, i.e. $\hat{\alpha}(y) \in \mathfrak{M} \times_\alpha H$ by Theorem 1.1. Let π be the isomorphism of $\mathfrak{M} \times_\alpha H$ onto $\mathfrak{M} \times_{\alpha^H} H$ satisfying $\pi(\alpha(x)) = \alpha^H(x)$ and $\pi(1 \otimes \rho(r)) = 1 \otimes \rho^H(r)$. Then

$$\alpha^H(\mathfrak{M}) \subset \pi(\mathcal{P}) \subset \mathfrak{M} \times_{\alpha^H} H . \tag{2.1}$$

Here we set $\delta = (\alpha^H)^\wedge$. Since $\mathfrak{M}$ is a factor, $\Gamma(\delta) = H$ by Theorem IV.1.5.

Now we claim that $\pi(\mathcal{P}) \times_\delta H$ is a factor. To this end, we need only to check the ergodicity of δ on the center of $\pi(\mathcal{P})$ by Theorem IV.3.3. Suppose that $\delta(z) = z \otimes 1_H$ for $z \in \pi(C_{\mathcal{P}})$. The inclusion relations (2.1) imply that

$$\alpha^H(\mathfrak{M}) \subset \pi(\mathcal{P})^\delta \subset (\mathfrak{M} \times_{\alpha^H} H)^\delta = \alpha^H(\mathfrak{M}) .$$

Since $\mathfrak{M}$ is a factor, so is $\pi(\mathcal{P})^\delta$. Since $\pi(C_{\mathcal{P}})^\delta$ is contained in the center of $\pi(\mathcal{P})^\delta$, z must be a scalar operator.

On the other hand, by (2.1), we have

$$\alpha^H(\mathfrak{M}) \times_\delta H \subset \pi(\mathfrak{P}) \times_\delta H \subset (\mathfrak{M} \times_{\alpha^H} H) \times_\delta H\ .$$

The duality for crossed product gives us an isomorphism π' of $(\mathfrak{M} \times_{\alpha^H} H) \times_\delta H$ onto $\mathfrak{M} \bar{\otimes} \mathcal{L}(L^2(H))$ as in the proof of Theorem I.2.7. Then $\pi'(\alpha^H(\mathfrak{M}) \times_\delta H) = \mathfrak{M} \bar{\otimes} L^\infty(H)$ by Lemma I.2.6. Therefore,

$$\mathfrak{M} \bar{\otimes} L^\infty(H) \subset \pi'(\pi(\mathfrak{P}) \times_\delta H) \subset \mathfrak{M} \bar{\otimes} \mathcal{L}(L^2(G))\ .$$

Since $\pi(\mathfrak{P}) \times_\delta H$ is a factor, it must be $\mathfrak{M} \bar{\otimes} \mathcal{L}(L^2(G))$. This means $\pi(\mathfrak{P}) \times_\delta H = (\mathfrak{M} \times_{\alpha^H} H) \times_\delta H$. Considering the fixed points of $\hat{\delta}$, we have $\pi(\mathfrak{P}) = \mathfrak{M} \times_{\alpha^H} H$ and hence $\mathfrak{P} = \mathfrak{M} \times_\alpha H$. Q.E.D.

An action $\hat{\delta}$ of G on $\mathfrak{N} \times_\delta G$ carries the generators $\delta(y)$ and $1 \otimes f$ to $\delta(y) \otimes 1$ and $1 \otimes \lambda(f)$, respectively, i.e. $\hat{\delta}$ leaves $\delta(\mathfrak{N})$ fixed but $\mathbf{C} \otimes L^\infty(G)$ not. On the contrary, we want to consider a co-action δ^d of G on $\mathfrak{N} \times_\delta G$ which makes $\mathbf{C} \otimes L^\infty(G)$ fix and $\delta(\mathfrak{N})$ move:

$$\delta(y) \mapsto (\iota \otimes \delta_G)(\delta(y)), \qquad 1 \otimes f \mapsto 1 \otimes f \otimes 1$$

which agree with a co-action of G on $\mathfrak{N} \times_\delta G$ defined by the map: $x \mapsto \mathrm{Ad}_{1 \otimes W_G^*}(x \otimes 1)$.

<u>Theorem</u> 2.2. Let $\mathfrak{P}$ be a von Neumann subalgebra of $\mathfrak{N} \times_\delta G$ containing $\delta(\mathfrak{N})$. If $\mathfrak{N} \times_\delta G$ is a factor, then the following two conditions are equivalent:

(i) $\mathfrak{P}$ is globally δ^d invariant.

(ii) $\mathfrak{P} = \mathfrak{N} \times_\delta (H \backslash G)$ for some closed subgroup H of G.

<u>Proof</u>. Without any loss of generality we may assume that $\mathfrak{N}$ is standard.

(i) $\Rightarrow$ (ii): Let ε be the co-action of G on $\delta(\mathfrak{N})'$ defined by (II.3.4). Then, by Corollary II.3.3 there exist an action α of G on $(\mathfrak{N} \times_\delta G)'$ and an isomorphism π of $\delta(\mathfrak{N})'$ onto $(\mathfrak{N} \times_\delta G)' \times_\alpha G$ such that $\hat{\alpha} \circ \pi = (\pi \otimes \iota) \circ \varepsilon$. We want to consider the following correspondences:

$$\begin{array}{ccccc}
\delta(\mathfrak{N}) & \longleftrightarrow & \{\delta(\mathfrak{N})', \varepsilon\} & \overset{\pi}{\to} & \{(\mathfrak{N} \times_\delta G)' \times_\alpha G, \hat{\alpha}\} \\
| & & | & & | \\
\mathfrak{P} & \longleftrightarrow & \mathfrak{P}' & \longrightarrow & \pi(\mathfrak{P}') \\
| & & | & & | \\
\mathfrak{N} \times_\delta G & \longleftrightarrow & (\mathfrak{N} \times_\delta G)' & \to & \alpha((\mathfrak{N} \times_\delta G)')
\end{array}\ .$$

Since $\mathfrak{P}$ is globally δ^d invariant, $\mathrm{Ad}_{1 \otimes W_G^*}(\mathfrak{P} \otimes \mathbf{C})$ is contained in $\mathfrak{P} \bar{\otimes} \mathcal{R}(G)$. Using the property $W_G \in L^\infty(G) \bar{\otimes} \mathcal{R}(G)$, we find that $\mathrm{Ad}_{1 \otimes W_G}(\mathfrak{P}' \otimes \mathbf{C})$ is contained in $\mathfrak{P}' \bar{\otimes} \mathcal{R}(G)$. This means that $\mathfrak{P}'$ is globally ε invariant, or $\pi(\mathfrak{P}')$ is globally $\hat{\alpha}$ invariant. Since $(\mathfrak{N} \times_\delta G)'$ is a factor, we can apply Theorem 2.1 and obtain that

$$\pi(\mathcal{P}') = (\mathfrak{n} \times_\delta G)' \times_\alpha H$$

for some closed subgroup H of G. Remembering the property (II.2.4) of π, we have

$$\pi(y) = \alpha(y), \qquad y \in (\mathfrak{n} \times_\delta G)'$$

$$\pi(1 \otimes \lambda(r)) = 1 \otimes 1 \otimes \rho(r), \quad r \in G .$$

Therefore, $\mathcal{P}'$ is generated $(\mathfrak{n} \times_\delta G)'$ and $\mathbf{C} \otimes \lambda(H)$, in other words, $\mathcal{P} = (\mathfrak{n} \times_\delta G) \cap (\mathbf{C} \otimes \lambda(H))' = \mathfrak{n} \times_\delta (H\backslash G)$ by Theorem 1.2.

(ii) $\Rightarrow$ (i): The commutant of $\mathcal{P} = \mathfrak{n} \times_\delta (H\backslash G)$ is generated by $(\mathfrak{n} \times_\delta G)'$ and $\mathbf{C} \otimes \lambda(H)$. Since $\varepsilon(y) = y \otimes 1$ for $y \in (\mathfrak{n} \times_\delta G)'$ and $\varepsilon(1 \otimes \lambda(r)) = 1 \otimes \lambda(r) \otimes \rho(r)$, it follows that $\mathcal{P}'$ is globally ε invariant. Therefore, $\mathrm{Ad}_{1\otimes W_G}(\mathcal{P}' \otimes \mathbf{C})$ is contained in $\mathcal{P}' \bar{\otimes} \mathcal{R}(G)$. Since $W_G \in L^\infty(G) \bar{\otimes} \mathcal{R}(G)$, $\mathcal{P}$ is globally δ^d invariant.

Q.E.D.

For a discrete group G we have a more general result.

Theorem 2.3. Assume that G is discrete. Let $\mathcal{P}$ be a von Neumann subalgebra of $\mathfrak{m} \times_\alpha G$ containing $\alpha(\mathfrak{m})$. If α is free and there exists a subgroup $\mathcal{S}$ of $\mathrm{Aut}(\mathfrak{m} \times_\alpha G)$ such that

(a) $\alpha(\mathfrak{m})$ is globally $\mathcal{S}$ invariant;

(b) $\mathcal{S}$ is ergodic on the center of $\alpha(\mathfrak{m})$;

(c) $\mathcal{S}$ is trivial on $\mathbf{C} \otimes \mathcal{R}(G)$,

then the following two conditions are equivalent:

(i) $\mathcal{P}$ is globally $\mathcal{S}$ invariant and of the form $\mathcal{E}(\mathfrak{m} \times_\alpha G)$ for some faithful normal expectation $\mathcal{E}$.

(ii) $\mathcal{P} = \mathfrak{m} \times_\alpha H$ for some subgroup H of G.

Proof. (ii) $\Rightarrow$ (i): The $\mathcal{S}$ invariance of $\mathfrak{m} \times_\alpha H$ is clear from (a) and (c). Let e_{tH} be the projection in $\ell^\infty(G)$ with support tH. Let $\mathcal{E}$ be a faithful normal expectation defined by

$$\mathcal{E}(y) = \sum_{tH \in G/H} (1 \otimes e_{tH}) y (1 \otimes e_{tH}), \qquad y \in \mathfrak{m} \times_\alpha G .$$

Since $e_{tH}\rho_r(e_{tH}) = e_{tH}$ for $r \in H$ and $e_{tH}\rho_r(e_{tH}) = 0$ for $r \notin H$, we have $\mathcal{E}(\mathfrak{m} \times_\alpha G) = \mathfrak{m} \times_\alpha H$.

(i) $\Rightarrow$ (ii): Since α is free, $\mathcal{E}(1 \otimes \rho(t)) = (1 \otimes \rho(t))e$ for some central projection e in $\mathcal{P}$ by Lemma 2.4 below. Since α is free, we have

$$\mathcal{P}' \cap \mathcal{P} \subset \alpha(\mathfrak{m})' \cap (\mathfrak{m} \times_\alpha G) = \alpha(\mathbf{C}_\mathfrak{m})$$

and hence $e \in \alpha(\mathbf{C}_\mathfrak{m})$. Since $\mathcal{P}$ is globally $\mathcal{S}$ invariant and α is free, we have $\mathcal{E} \circ \gamma = \gamma \circ \mathcal{E}$. Since $\mathcal{S}$ is a trivial on $\mathbf{C} \otimes \mathcal{R}(G)$, we have

$$(1 \otimes \rho(t))e = (1 \otimes \rho(t))\gamma(e), \qquad \gamma \in \mathcal{S} .$$

Therefore the ergodicity of $\mathfrak{g}$ on $\alpha(C_{\mathfrak{m}})$ implies $e = 0$ or 1. Let H be the set of $t \in G$ with $\mathcal{E}(1 \otimes \rho(t)) = 1 \otimes \rho(t)$. Then H is a subgroup of G and $\mathfrak{P} = \mathcal{E}(\mathfrak{m} \times_{\alpha} G) = \mathfrak{m} \times_{\alpha} H$. Q.E.D.

Lemma 2.4. Let $\mathfrak{n}$ be a von Neumann subalgebra of $\mathfrak{m}$ with $\mathfrak{n}' \cap \mathfrak{m} \subset \mathfrak{n}$ and $\mathcal{E}$ a faithful normal expectation of $\mathfrak{m}$ onto $\mathfrak{n}$. If u is a unitary in $\mathfrak{m}$ with $u\mathfrak{n}u^* = \mathfrak{n}$, then $\mathcal{E}(u) = ue = eu$ for some central projection e in $\mathfrak{n}$.

Proof. Since $\mathcal{E}(u)x = \mathcal{E}(uxu^*u) = uxu^*\mathcal{E}(u)$ for all $x \in \mathfrak{n}$, we have $u^*\mathcal{E}(u) \in C_{\mathfrak{n}}$ and hence

$$\mathcal{E}(u)\mathcal{E}(u)^*\mathcal{E}(u) = \mathcal{E}(u)u^*\mathcal{E}(u) = \mathcal{E}(uu^*\mathcal{E}(u)) = \mathcal{E}(u) .$$

Therefore $\mathcal{E}(u)$ is a partial isometry. We set

$$e = \mathcal{E}(u)^*\mathcal{E}(u) = \mathcal{E}(u^*\mathcal{E}(u)) = u^*\mathcal{E}(u) .$$

Then $e \in C_{\mathfrak{n}}$ and $\mathcal{E}(u) = ue = eu$, for $u^*\mathcal{E}(u)x = xu^*\mathcal{E}(u)$ for $x \in \mathfrak{n}$. Q.E.D.

NOTES

Theorem 2.1 is obtained by [69] for abelian locally compact groups and [47] for non abelian ones. Theorem 2.2 is obtained in [47]. Theorem 2.3 is obtained by H. Choda [6].

§3. Galois correspondences

Making use of the results obtained in §§1 and 2, we shall give a Galois correspondence. For a von Neumann algebra $\mathcal{P}_1$ and its von Neumann subalgebra $\mathcal{P}_2$ we denote by $\mathcal{R}(\mathcal{P}_1,\mathcal{P}_2)$ the set of all von Neumann subalgebras $\mathcal{P}$ of $\mathcal{P}_1$ containing $\mathcal{P}_2$.

Theorem 3.1. If $\mathcal{N}^\delta$ is a factor and δ is semi-dual, then there exists a bijective correspondence between the set of closed subgroups H of G and the set of globally δ invariant von Neumann subalgebras $\mathcal{P} \in \mathcal{R}(\mathcal{N},\mathcal{N}^\delta)$:

$$(3.1)\qquad \begin{aligned} H &\mapsto \mathcal{P}(H) = \{y \in \mathcal{N} : \delta(y) \in \mathcal{N} \bar{\otimes} \rho(H)''\} , \\ \mathcal{P} &\mapsto H(\mathcal{P}) = \bigcap \{K : \delta(\mathcal{P}) \subset \mathcal{N} \bar{\otimes} \rho(K)''\} , \end{aligned}$$

where K runs over all closed subgroups of G.

Proof. The case where $\mathcal{N}^\delta$ is properly infinite: By virtue of Theorem III.4.4, δ is dominant on $\mathcal{N}$. Combining Theorems 1.1 and 2.1, we have the desired correspondence.

The general case: Put $\bar{\mathcal{N}} = \mathcal{N} \bar{\otimes} F_\infty$, $\bar{\delta} = (\iota \otimes \sigma) \circ (\delta \otimes \iota)$ and $\bar{\mathcal{P}} = \mathcal{P} \bar{\otimes} F_\infty$. Then $\bar{\mathcal{N}}^{\bar{\delta}} = \mathcal{N}^\delta \bar{\otimes} F_\infty$ is a properly infinite factor and $\bar{\delta}$ is semi-dual. Therefore

$$H = H(\bar{\mathcal{P}}(H)) \quad \text{and} \quad \bar{\mathcal{P}} = \bar{\mathcal{P}}(H(\bar{\mathcal{P}})) ,$$

where $\bar{\mathcal{P}}(H)$ and $H(\bar{\mathcal{P}})$ are defined similarly as (3.1). Since $\bar{\mathcal{P}}(H) = \mathcal{P}(H) \bar{\otimes} F_\infty$, we have $H = H(\bar{\mathcal{P}}(H)) = H(\mathcal{P}(H) \bar{\otimes} F_\infty) = H(\mathcal{P}(H))$. Since $\delta(\mathcal{P}) \subset \mathcal{N} \bar{\otimes} \rho(H)''$ is equivalent to $\bar{\delta}(\bar{\mathcal{P}}) \subset \bar{\mathcal{N}} \bar{\otimes} \rho(H)''$, we have $H(\mathcal{P}) = H(\bar{\mathcal{P}})$. Therefore, if $x \in \mathcal{P}(H(\mathcal{P}))$, then $x \otimes 1 \in \bar{\mathcal{P}}(H(\bar{\mathcal{P}}))$ and hence $x \in \mathcal{P}$. Thus, $\mathcal{P}(H(\mathcal{P})) = \mathcal{P}$. Q.E.D.

Corollary 3.2. Assume that $\mathcal{N}$ or $\mathcal{N}^\delta$ is a factor. If δ is integrable and $\Gamma(\delta) = G$, then there exists the same correspondence as in Theorem 3.1.

Proof. The case where $\mathcal{N}^\delta$ is properly infinite: By Theorem IV.3.4, δ is dominant on $\mathcal{N}$. Therefore, $\mathcal{N}^\delta \bar{\otimes} \mathcal{L}(L^2(G)) \cong \mathcal{N} \times_\delta G$. Since $\Gamma(\delta) = G$ and $\mathcal{N}$ is a factor, $\mathcal{N} \times_\delta G$ is a factor by Theorem IV.3.3. Therefore, $\mathcal{N}^\delta$ is a factor. The remainder of the proof is the same as in Theorem 3.1. Q.E.D.

Theorem 3.3. If $\mathcal{M}$ is a factor and α is semi-dual, then there exist a co-action δ^α of G on $\mathcal{M}$ and a bijective correspondence between the set of closed subgroups H of G and the set of globally δ^α invariant von Neumann subalgebras $\mathcal{P} \in \mathcal{R}(\mathcal{M},\mathcal{M}^\alpha)$ in such a way that

$$(3.2)\qquad \begin{aligned} H &\mapsto \mathcal{P}(H) = \{x \in \mathcal{M} : \alpha_t(x) = x, \ t \in H\} , \\ \mathcal{P} &\mapsto H(\mathcal{P}) = \{t \in G : \alpha_t(x) = x, \ x \in \mathcal{P}\} . \end{aligned}$$

Proof. The case where $\mathfrak{M}^{\alpha}$ is properly infinite: Since α is dominant, Theorems 1.2 and 2.2 give us the desired correspondence.

The general case: Put $\bar{\mathfrak{M}} = \mathfrak{M} \bar{\otimes} F_{\infty}$, $\bar{\alpha}_t = \alpha_t \otimes \iota$ and $\bar{\mathfrak{P}} = \mathfrak{P} \bar{\otimes} F_{\infty}$. Then $\bar{\mathfrak{M}}$ is a factor, $\bar{\alpha}$ is semi-dual and $\bar{\mathfrak{M}}^{\bar{\alpha}} = \mathfrak{M}^{\alpha} \bar{\otimes} F_{\infty}$ is properly infinite. Therefore

$$H = H(\bar{\mathfrak{P}}(H)) \quad \text{and} \quad \bar{\mathfrak{P}} = \bar{\mathfrak{P}}(H(\bar{\mathfrak{P}})) \ ,$$

where $\bar{\mathfrak{P}}(H)$ and $H(\bar{\mathfrak{P}})$ are defined similarly as (3.2). If $t \in H(\mathfrak{P}(H))$, then $\alpha_t = \iota$ on $\mathfrak{P}(H)$. Since $\bar{\mathfrak{P}}(H) = \mathfrak{P}(H) \bar{\otimes} F_{\infty}$. it follows that $\bar{\alpha}_t = \iota$ on $\bar{\mathfrak{P}}(H)$ and hence $t \in H(\bar{\mathfrak{P}}(H)) = H$. Thus $H(\mathfrak{P}(H)) = H$. If $x \in \mathfrak{P}(H(\mathfrak{P}))$, then $\alpha_t(x) = x$ for $t \in H(\mathfrak{P})$ and hence $\bar{\alpha}_t(x \otimes 1) = x \otimes 1$ for $t \in H(\mathfrak{P})$. Since $H(\mathfrak{P}) = H(\bar{\mathfrak{P}})$, it follows that $x \otimes 1 \in \bar{\mathfrak{P}}(H(\bar{\mathfrak{P}})) = \bar{\mathfrak{P}}$ and hence $x \in \mathfrak{P}$. Thus, $\mathfrak{P}(H(\mathfrak{P})) = \mathfrak{P}$. Q.E.D.

Corollary 3.4. Assume that $\mathfrak{M}$ is a factor. If α is integrable and $\mathcal{C}_{\mathfrak{M} \times_{\alpha} G} \subset \alpha(\mathfrak{M})$, then there exists a co-action δ^{α} of G on $\mathfrak{M}$ and the same correspondence as in Theorem 3.3.

Proof. By Theorems 3.3 and IV.3.4. Q.E.D.

Lemma 3.5. Assume that G is discrete. If α is a free action implemented by a unitary representation u, then the following two conditions are equivalent:

(i) There exists a faithful normal expectation of $(\mathfrak{M}^{\alpha})'$ onto $\mathfrak{M}'$.

(ii) There exists an action γ of G on $\mathfrak{M}'$ and an isomorphism π of $(\mathfrak{M}^{\alpha})'$ onto $\mathfrak{M}' \times_{\gamma} G$ such that $\pi(x) = \gamma(x)$ for $x \in \mathfrak{M}'$ and $\pi(u(t)) = 1 \otimes \rho(t)$ for t.

Proof. Put $\gamma_t(x) = u(t)xu(t)^*$ for $x \in \mathfrak{M}'$. Then $\mathcal{E}(u(t)^*)\gamma_t(x) = x\mathcal{E}(u(t)^*)$ for $x \in \mathfrak{M}'$. Since α is free, so is γ. Thus $\mathcal{E}(u(t)) = 0$ for $t \neq e$. Let φ be a faithful normal state on $\mathfrak{M}'$, $\mathcal{E}$ a faithful normal expectation of $(\mathfrak{M}^{\alpha})'$ onto $\mathfrak{M}'$ and $\psi = \varphi \circ \mathcal{E}$. Let $\{\pi_{\varphi}, \mathfrak{H}_{\varphi}, \xi_{\varphi}\}$ and $\{\pi_{\psi}, \mathfrak{H}_{\psi}, \xi_{\psi}\}$ be GNS representations of $\mathfrak{M}'$ and $(\mathfrak{M}^{\alpha})'$, respectively. Since, for $x(t), y(t) \in \mathfrak{M}'$,

$$\left(\sum_t u(t)^* x(t) \Big| \sum_s u(s)^* y(s)\right)_{\psi} = \psi\left(\sum_{t,s} y(s)^* u(st^{-1}) x(t)\right)$$
$$= \varphi\left(\sum_{t,s} y(s)^* \mathcal{E}(u(st^{-1})) x(t)\right) = \sum (x(s) | y(s))_{\varphi} \ ,$$

there exists an isometry v of $\mathfrak{H}_{\psi}$ onto $\mathfrak{H}_{\varphi} \otimes \ell^2(G)$ defined by

$$v : \pi_{\psi}\left(\sum_s u(s)^* y(s)\right)\xi_{\psi} \mapsto \sum_s \pi_{\varphi}(y(s))\xi_{\varphi} \otimes f_s \ ,$$

where f_s is a function in $\ell^2(G)$ with support $\{s\}$ and $\|f_s\|_2 = 1$. By direct computation we have

$$v\pi_\psi(x)v^{-1} = (\pi_\varphi \otimes \iota)(\gamma(x)) , \quad x \in \mathfrak{M}' ,$$
$$v\pi_\psi(u(r))v^{-1} = 1 \otimes \rho(r) , \quad r \in G .$$

Setting $\pi = (\pi_\varphi \otimes \iota)^{-1} \circ \mathrm{Ad}_v \circ \pi_\psi$, we have the desired result. Q.E.D.

When G is discrete and α is semi-dual, then there exists a faithful normal expectation of $(\mathfrak{M}^\alpha)'$ onto $\mathfrak{M}'$. Indeed, we have only to consider the commutant of the spatial equivalence $\{\overline{\mathfrak{M}},\overline{\alpha}\} \cong \{\overline{\mathfrak{M}},\tilde{\alpha}\}$.

Theorem 3.6. Assume that G is discrete and there exists a faithful normal expectation of $(\mathfrak{M}^\alpha)'$ onto $\mathfrak{M}'$. If α is free and there is a subgroup $\mathfrak{S}$ of $\mathrm{Aut}(\mathfrak{M})$ satisfying

(i) $\mathfrak{S}$ is ergodic on $\mathcal{C}_{\mathfrak{M}}$; and

(ii) $\gamma(u(t)) = u(t)$ for all $\gamma \in \mathfrak{S}$ and $t \in G$,

then there exists a bijective correspondence between the set of subgroups H of G and the set of globally $\mathfrak{S}$ invariant von Neumann subalgebras $\mathfrak{P} \in \mathcal{R}(\mathfrak{M},\mathfrak{M}^\alpha)$ with a faithful normal expectation of $(\mathfrak{M}^\alpha)'$ onto $\mathfrak{P}'$ in such a way as (3.2).

Proof. Since α is free and there exists a faithful normal expectation of $(\mathfrak{M}^\alpha)'$ onto $\mathfrak{M}'$, there exist an action γ of G on $\mathfrak{M}'$ and an isomorphism π of $(\mathfrak{M}^\alpha)'$ onto $\mathfrak{M}' \times_\gamma G$ such that $\pi(\mathfrak{M}') = \gamma(\mathfrak{M}')$ and $\pi(u(t)) = 1 \otimes \rho(t)$, $t \in G$. Since γ is free and there exists a faithful normal expectation of $\mathfrak{M}' \times_\gamma G$ onto $\pi(\mathfrak{P}')$ we can apply Theorem 2.3 to $\{\mathfrak{M}',\gamma\}$. We consider the diagram:

$$\begin{array}{ccccccc}
\mathfrak{M}^\alpha & \longleftrightarrow & (\mathfrak{M}^\alpha)' & \xrightarrow{\pi} & \mathfrak{M}' \times_\gamma G & \longleftrightarrow & \{e\} \\
| & & | & & | & & | \\
\mathfrak{P} & \longleftrightarrow & \mathfrak{P}' & \longrightarrow & \pi(\mathfrak{P}') = \mathfrak{M}' \times_\gamma K & \longleftrightarrow & K \\
| & & | & & | & & | \\
\mathfrak{M} & \longleftrightarrow & \mathfrak{M}' & \longrightarrow & \gamma(\mathfrak{M}') & \longleftrightarrow & G .
\end{array}$$

Since $\alpha_t = \iota$ on $\mathfrak{P}$ if and only if $1 \otimes \rho(t) \in \pi(\mathfrak{P}')$, we have $H(\mathfrak{P}) = K(\pi(\mathfrak{P}'))$. Since $\mathfrak{D}(K(\mathfrak{D})) = \mathfrak{D}$ $(\mathfrak{D} = \pi(\mathfrak{P}'))$, by Theorem 2.3, we have $\pi(\mathfrak{P}') = \mathfrak{M}' \times_\gamma H(\mathfrak{P})$. This implies $\mathfrak{P}' = \mathfrak{M}' \vee u(H(\mathfrak{P}))''$ and so $\mathfrak{P} = \mathfrak{P}(H(\mathfrak{P}))$. Since $\mathfrak{P}(H)$ corresponds to $\mathfrak{M}' \times_\gamma H$ in the diagram, we have $t \in H(\mathfrak{P}(H))$ is equivalent to $1 \otimes \rho(t) \in \mathfrak{M}' \times_\gamma H$. This shows that $H(\mathfrak{P}(H)) = H$. Q.E.D.

Remark. Up to this point, we often use the following correspondences:

(a) If $\{\mathfrak{M},\alpha\} = \{\mathfrak{N} \times_\delta G, \hat{\delta}\}$ and ε the co-action of G on $\delta(\mathfrak{N})'$ defined by (II.3.4), then there exists an action γ of G on $(\mathfrak{N} \times_\delta G)'$ and an isomorphism of $\delta(\mathfrak{N})'$ onto $(\mathfrak{N} \times_\delta G)' \times_\gamma G$ such that

$$\begin{array}{ccccc}
\mathfrak{M} = \mathfrak{N} \times_\delta G & \longleftrightarrow & (\mathfrak{N} \times_\delta G)' & \cong & \gamma((\mathfrak{N} \times_\delta G)') \\
\downarrow \mathcal{E}_{11} & & | & & \uparrow \mathcal{E}_{22} \\
\mathfrak{M}^{\alpha^H} = \mathfrak{N} \times_\delta (H\backslash G) & \longleftrightarrow & (\mathfrak{N} \times_\delta (H\backslash G))' & \cong & (\mathfrak{N} \times_\delta G)' \times_\gamma H \\
\downarrow \mathcal{E}_{12} & & | & & \uparrow \mathcal{E}_{21} \\
\mathfrak{M}^\alpha = \delta(\mathfrak{N}) & \longleftrightarrow & \{\delta(\mathfrak{N})', \varepsilon\} & \cong & \{(\mathfrak{N} \times_\delta G)' \times_\gamma G, \hat{\gamma}\} .
\end{array}$$

Here, $\mathcal{E}_{ij}$ are operator valued faithful semi-finite normal weights. Further, $\mathcal{E}_{11}$ and $\mathcal{E}_{12}$ are bounded for a compact G, and $\mathcal{E}_{21}$ and $\mathcal{E}_{22}$ are bounded for a discrete G.

(b) If $\{\mathfrak{N},\delta\} = \{\mathfrak{M} \times_\alpha G, \hat{\alpha}\}$ and β the action of G on $\alpha(\mathfrak{M})'$ defined by (II.3.3), then there exists a co-action ζ of G on $(\mathfrak{M} \times_\alpha G)'$ and an isomorphism of $\alpha(\mathfrak{M})'$ onto $(\mathfrak{M} \times_\alpha G)' \times_\zeta G$ such that

$$\begin{array}{ccccc}
\mathfrak{N} = \mathfrak{M} \times_\alpha G & \longleftrightarrow & (\mathfrak{M} \times_\alpha G)' & \cong & \zeta((\mathfrak{M} \times_\alpha G)') \\
\downarrow \mathcal{E}_{41} & & | & & \uparrow \mathcal{E}_{32} \\
\mathfrak{M} \times_\alpha H & \longleftrightarrow & (\mathfrak{M} \times_\alpha H)' & \cong & (\mathfrak{M} \times_\alpha G)' \times_\zeta (H\backslash G) \\
\downarrow \mathcal{E}_{42} & & | & & \uparrow \mathcal{E}_{31} \\
\mathfrak{N}^\delta = \alpha(\mathfrak{M}) & \longleftrightarrow & \{\alpha(\mathfrak{M})', \beta\} & \cong & \{(\mathfrak{M} \times_\alpha G)' \times_\zeta G, \hat{\zeta}\} .
\end{array}$$

Here, $\mathcal{E}_{ij}$ are operator valued faithful semi-finite normal weights. Further, $\mathcal{E}_{31}$ and $\mathcal{E}_{32}$ are bounded for a compact G, and $\mathcal{E}_{41}$ and $\mathcal{E}_{42}$ are bounded for a discrete G.

Theorem 3.7. Assume that G is compact. If α is faithful and there is an ergodic subgroup $\mathcal{S}$ of $\mathrm{Aut}(\mathfrak{M})$ commuting with α_t, $t \in G$, then there exists a bijective correspondence of (3.2) between the set of closed normal subgroups H of G and the set of globally α and $\mathcal{S}$ invariant von Neumann sublagebras $\mathcal{P} \in \mathcal{R}(\mathfrak{M},\mathfrak{M}^\alpha)$.

Proof. The case where $\mathfrak{M}^\alpha$ is properly infinite: By means of Proposition IV.2.5, α is dominant on $\mathfrak{M}$. Therefore, $H = H(\mathcal{P}(H))$ by Theorem 1.2. It suffices to show that $\mathcal{P}(H(\mathcal{P}))$. It is clear that $\mathcal{P} \subset \mathcal{P}(H(\mathcal{P}))$. Since $H(\mathcal{P})$ is normal, α is considered as a faithful action of $G/H(\mathcal{P})$ on $\mathcal{P}(H(\mathcal{P}))$. Since the fixed points are $\mathfrak{M}^\alpha$, we may assume that $H(\mathcal{P}) = \{e\}$ and must show that $\mathcal{P} = \mathfrak{M}$.

From this assumption, α is faithful on $\mathcal{P}$ and hence α is dominant on $\mathcal{P}$, because $\mathcal{P}$ is globally α and $\mathcal{S}$ invariant. By means of Theorem II.2.2, there exists an isomorphism π of $L^\infty(G)$ into $\mathcal{P}$ with $\alpha_t \circ \pi = \pi \circ \lambda_t$ and $\mathcal{P}$ is generated by $\mathfrak{M}^\alpha = \mathcal{P}^\alpha$ and $\pi(L^\infty(G))$. Since π is also an isomorphism of $L^\infty(G)$ into $\mathfrak{M}$ with $\alpha_t \circ \pi = \pi \circ \lambda_t$, $\mathfrak{M}$ is generated by $\mathfrak{M}^\alpha$ and $\pi(L^\infty(G))$. Thus $\mathcal{P} = \mathfrak{M}$.

The general case: Put $\bar{\mathfrak{M}} = \mathfrak{M} \bar{\otimes} F_\infty$, $\bar{\alpha}_t = \alpha_t \otimes \iota$, $\bar{\mathcal{P}} = \mathcal{P} \bar{\otimes} F_\infty$ and

$\bar{\mathcal{S}} = \{\tau \otimes \tau' : \tau \in \mathcal{S},\ \tau' \in \mathrm{Aut}(F_\infty)\}$. Then

$$H = H(\bar{\mathcal{P}}(G)) \quad \text{and} \quad \bar{\mathcal{P}} = \bar{\mathcal{P}}(H(\bar{\mathcal{P}})) .$$

The rest of the proof is the same as that of Theorem 3.3. Q.E.D.

Theorem 3.8. Assume that G is compact. If there is an ergodic subgroup $\mathcal{S}$ of $\mathrm{Aut}(\mathfrak{M})$ commuting with α_t, $t \in G$, then $\sigma \in \mathrm{Aut}(\mathfrak{M}/\mathfrak{M}^\alpha)$ commuting with $\mathcal{S}$ is of the form $\sigma = \alpha_r$ for some $r \in G$.

Proof. The case where $\mathfrak{M}^\alpha$ is properly infinite: By virtue of Propositions IV.2.2 and IV.2.4, every irreducible subrepresentation of $\{\alpha, \mathfrak{M}\}$ is equivalent to some representation in $\mathcal{H}_\alpha(\mathfrak{M})$. Therefore, Theorem I.3.4 tells us that we need only to check

$$(3.3) \qquad \sigma(\mathfrak{K}) = \mathfrak{K} \quad \text{for all} \quad \mathfrak{K} \in \mathcal{H}_\alpha(\mathfrak{M}) .$$

Let $\{v_j : j = 1, \cdots, d\}$ be an orthonormal basis of $\mathfrak{K} \in \mathcal{H}_\alpha(\mathfrak{M})$. Let $\mathfrak{K}_0$ be a Hilbert space in $\mathfrak{M}^\alpha$ with $\dim \mathfrak{K}_0 = d$ and $\{u_j : j = 1, \cdots, d\}$ its orthonormal basis. We set

$$w = \sum_{j=1}^{d} v_j u_j^* \quad \text{and} \quad V(t) = \sum_{j,k} v_k^* \alpha_t(v_j) u_k u_j^* .$$

Then w is an isometry of $\mathfrak{K}_0$ onto $\mathfrak{K}$ and $\{V, \mathfrak{K}_0\} = w^*\{\alpha, \mathfrak{K}\}w$. Indeed,

$$(3.4) \qquad \alpha_t(w) = \sum \alpha_t(v_j) u_j^* = wV(t) .$$

Since each element in $\mathfrak{M}$ is a linear span of elements of the form

$$\sum \lambda_{jk} u_j x u_k^* = \rho_0(x) \sum \lambda_{jk} u_j u_k^* ; \qquad \lambda_{jk} \in \mathbf{C}, \quad x \in \mathfrak{M} ,$$

$\mathfrak{M}$ is considered as $\rho_0(\mathfrak{M}) \bar{\otimes} (\mathfrak{K}_0, \mathfrak{K}_0)$, where ρ_0 is the endomorphism of $\mathfrak{M}$ defined by $\rho_0(x) = \sum u_j x u_j^*$ and $(\mathfrak{K}_0, \mathfrak{K}_0)$ is the subspace of those elements in $\mathfrak{M}$ mapping $\mathfrak{K}_0$ into itself. Since $(\mathfrak{K}_0, \mathfrak{K}_0)$ is contained in $\mathfrak{M}^\alpha$, it follows that

$$\alpha_t = \alpha_t \otimes \iota \quad \text{and} \quad \sigma = \sigma \otimes \iota$$

on $\mathfrak{M} = \rho_0(\mathfrak{M}) \bar{\otimes} (\mathfrak{K}_0, \mathfrak{K}_0)$. For each $\tau \in \mathcal{S}$ we define τ_0 and $\bar{\tau}$ by

$$\tau_0 \circ \rho_0 = \rho_0 \circ \tau \quad \text{and} \quad \bar{\tau} = \tau_0 \otimes \iota$$

on $\mathfrak{M} = \rho_0(\mathfrak{M}) \bar{\otimes} (\mathfrak{K}_0, \mathfrak{K}_0)$. Since $\mathcal{S}$ is ergodic on $\mathfrak{M}$, so is $\{\tau_0 : \tau \in \mathcal{S}\}$ on $\rho_0(\mathfrak{M})$. The proof of (3.3) is now obtained as follows. Since $\mathfrak{K} = w\mathfrak{K}_0$, the inclusion $\sigma(\mathfrak{K}) \subset \mathfrak{K}$ is equivalent to the fact that $w^*\sigma(w) \in (\mathfrak{K}_0, \mathfrak{K}_0)$. But, since $\{\tau_0 : \tau \in \mathcal{S}\}$ is ergodic on $\rho_0(\mathfrak{M})$, it suffices to check

$$(3.5) \qquad \bar{\tau}(w^*\sigma(w)) = w^*\sigma(w) .$$

To this end we set $a_\tau = \bar{\tau}(w)w^*$. Since

$$\alpha_t(a_\tau) = \bar{\tau}(wV(t))V(t)^*w^* = a_\tau$$

by (3.4), it follows from the assumption on σ that $\sigma(a_\tau) = a_\tau$, which implies (3.5).

The general case: We set $\bar{\mathfrak{M}} = \mathfrak{M} \bar{\otimes} F_\infty$, $\bar{\alpha}_t = \alpha_t \otimes \iota$, $\bar{\sigma} = \sigma \otimes \iota$ and $\bar{\mathcal{G}} = \{\tau \otimes \tau' : \tau \in \mathcal{G}, \ \tau' \in \mathrm{Aut}(F_\infty)\}$. Then $\bar{\sigma} = \bar{\alpha}_r$ for some r. Thus, $\sigma = \alpha_r$.

Q.E.D.

NOTES

A Galois correspondence for continuous groups (Theorem 3.1) is obtained by Connes and Takesaki [14] for abelian locally compact groups and [47] for non abelian ones. Theorem 3.3 is obtained in [47]. Lemma 3.5 is obtained by Connes [12] and M. Choda [4]. A Galois correspondence for discrete groups (Theorem 3.6) is obtained by H. Choda [6]. A Galois correspondence for compact groups (Theorem 3.7) is obtained by Kishimoto [41] by a slightly different method. Theorem 3.8 is obtained by Araki, Kastler, Takesaki and Haag [3].

§4. Galois correspondences (II)

In this final section we shall give a different sort of Galois correspondence. In the first half of this section we introduce the concept of a full group of automorphisms of $\mathfrak{M}$ and get a Galois type correspondence finer than Theorem 2.3, which will be called the Dye correspondence. The second half is an introduction to a new approach to the Galois correspondence.

In what follows we assume that G is a (countable) discrete group and α a faithful action of G on $\mathfrak{M}$. We occasionally identify G with the subgroup $\{\alpha_t : t \in G\}$ of $\mathrm{Aut}(\mathfrak{M})$ and use the same symbol G for them.

1. Dye correspondence

For any β, $\gamma \in \mathrm{Aut}(\mathfrak{M})$ there exists the largest central projection $e = e(\beta,\gamma)$ satisfying that the automorphism $(\gamma^{-1}\beta)^e$ is inner on $\mathfrak{M}_e$.

Definition 4.1. For a faithful action α of G on $\mathfrak{M}$ the set of all $\beta \in \mathrm{Aut}(\mathfrak{M})$ satisfying

$$\sup_{t \in G} e(\beta,\alpha_t) = 1$$

is called the full group of G, which is denoted by $[G]$. A group G is called full if $G = [G]$.

For each α there exists a partition $\{e(t) : t \in G\}$ of the identity in $\mathcal{C}_{\mathfrak{M}}$ such that $\{\alpha_t^{-1}(e(t)) : t \in G\}$ is also a partition of the identity. Then $\beta \in [G]$ if and only if

$$(4.1) \qquad \beta(x) = \sum_{t \in G} e(t)\alpha_t(vxv^*) , \qquad x \in \mathfrak{M}$$

for some unitary $v \in \mathfrak{M}$. Therefore, $\beta \in [G]$ if and only if $\beta = \mathrm{Ad}_u \circ \alpha$ for some $u \in N(\alpha(\mathfrak{M}))$, where $N(\alpha(\mathfrak{M}))$ is the normalizer of $\alpha(\mathfrak{M})$ in $\mathfrak{M} \times_\alpha G$, i.e. the set of unitaries u in $\mathfrak{M} \times_\alpha G$ with $u\alpha(\mathfrak{M})u^* = \alpha(\mathfrak{M})$.

Theorem 4.2. (Dye correspondence) If α is free, there exists a bijective correspondence between the set of full subgroups H of $[G]$ and the set of von Neumann subalgebras $\mathfrak{Q} \in \mathfrak{R}(\mathfrak{M} \times_\alpha G, \alpha(\mathfrak{M}))$ with a faithful normal conditional expectation of $\mathfrak{M} \times_\alpha G$ onto $\mathfrak{Q}$:

$$\mathfrak{Q} \mapsto H(\mathfrak{Q}) = \{\gamma \in \mathrm{Aut}(\mathfrak{M}) : \alpha \circ \gamma = \mathrm{Ad}_u \circ \alpha, \ \ u \in N(\alpha(\mathfrak{M})) \cap \mathfrak{Q}\} ;$$

$$H \mapsto \mathfrak{Q}(H) = \{u \in N(\alpha(\mathfrak{M})) : \alpha \circ \gamma = \mathrm{Ad}_u \circ \alpha, \ \ \gamma \in H\}'' .$$

Proof. See [37].

In order to translate the above theorem into the Galois type correspondence, we choose a suitable Hilbert space $\mathfrak{H}$ so that the action α is implemented by a unitary representation u of G on $\mathfrak{H}$ with $\alpha_t = \mathrm{Ad}_{u(t)}$ on $\mathcal{M}$. Then α', defined by

$$(4.2) \qquad \alpha_t'(x) = u(t)xu(t)^*, \qquad x \in \mathcal{M}'$$

is an action of G on $\mathcal{M}'$. For each $\beta \in [G]$ of the form (4.1) we define β' by

$$\beta'(x) = \sum_{t \in G} e(t)\alpha_t'(x), \qquad x \in \mathcal{M}'$$

is an automorphism of $\mathcal{M}'$. Denote the set of all such β' with $\beta \in [G]$ by $[G]_C$ and call the C-<u>full</u> group determined by G.

<u>Theorem</u> 4.3. Assume that

(i) α is implemented by a unitary representation u of G;

(ii) α is free; and

(iii) there exists a faithful normal conditional expectation of $(\mathcal{M}^\alpha)'$ onto $\mathcal{M}'$.

If α' is defined by (4.2) and $[G]$ is the full group of $\{\alpha_t' : t \in G\}$ in $\mathrm{Aut}(\mathcal{M}')$, then there exists a bijective correspondence between the set of C-full subgroups H_C of $[G]_C$ and the set of von Neumann subalgebras $\mathcal{P} \in \mathcal{R}(\mathcal{M}, \mathcal{M}^\alpha)$ with a faithful normal expectation of $(\mathcal{M}^\alpha)'$ onto $\mathcal{P}'$:

$$(4.3) \qquad \begin{aligned} \mathcal{P} &\mapsto \{\mathrm{Ad}_u \restriction \mathcal{M} : u \in N(\mathcal{M}') \cap \mathcal{P}'\}, \\ H_C &\mapsto \mathcal{M} \cap \{u \in N(\mathcal{M}') : \mathrm{Ad}_u \in H_C\}', \end{aligned}$$

where $N(\mathcal{M}')$ is the normalizer of $\mathcal{M}'$ in $(\mathcal{M}^\alpha)'$.

<u>Proof</u>. According to our assumptions (i), (ii) and (iii), Lemma 3.5 gives an action γ $(= \alpha')$ of G on $\mathcal{M}'$ and an isomorphism π of $(\mathcal{M}^\alpha)'$ onto $\mathcal{M}' \times_\gamma G$ such that $\pi(x) = \gamma(x)$, $x \in \mathcal{M}'$ and $\pi(u(t)) = 1 \otimes \rho(t)$, $t \in G$. Since γ is free and there exists a faithful normal expectation of $\mathcal{M}' \times_\gamma G$ onto $\pi(\mathcal{P}')$ from our assumption on α and $\mathcal{P}$, we can apply Theorem 4.2 to $\{\mathcal{M}', \gamma\}$. Now, we consider the diagram:

$$\begin{array}{ccccccccc}
\mathcal{M}^\alpha & \longleftrightarrow & (\mathcal{M}^\alpha)' & \overset{\pi}{\rightarrow} & \mathcal{M}' \times_\gamma G & \longleftrightarrow & [G] & \longleftrightarrow & [G]_C \\
| & & | & & | & & | & & | \\
\mathcal{P} & \longleftrightarrow & \mathcal{P}' & \rightarrow & \pi(\mathcal{P}') & \longleftrightarrow & H & \longleftrightarrow & H_C \\
| & & | & & | & & | & & | \\
\mathcal{M} & \longleftrightarrow & \mathcal{M}' & \rightarrow & \gamma(\mathcal{M}') & \longleftrightarrow & \mathrm{Int}(\mathcal{M}') & \longleftrightarrow & \{\iota\}
\end{array}.$$

Since $\pi(x) = \gamma(x)$, $x \in \mathcal{M}'$ and $\pi(N(\mathcal{M}')) = N(\gamma(\mathcal{M}'))$, if $\theta \in \mathrm{Aut}(\mathcal{M}')$ and $u \in N(\gamma(\mathcal{M}'))$, then $\gamma \circ \theta = \mathrm{Ad}_u \circ \gamma$ is equivalent to $\theta = \mathrm{Ad}_{\pi^{-1}(u)}$. Therefore, Theorem 4.2 gives the correspondence between $\mathcal{P}'$ and H:

$$H \mapsto \mathfrak{M} \vee \{u \in N(\mathfrak{M}') : Ad_u \in H\}'' ,$$

$$\mathcal{P}' \mapsto \{Ad_u \restriction \mathfrak{M}' : u \in N(\mathfrak{M}') \cap \mathcal{P}'\} ,$$

which implies our result. Q.E.D.

Remark. If α is free, then there exists an injection: $H \mapsto [H]$ from the set of subgroups of G into the set of full subgroups of $[G]$. This injection is not necessarily surjective. Moreover, $\mathfrak{M} \times_\alpha H = \mathfrak{D}([H])$. Indeed, $\mathfrak{M} \times_\alpha H \subset \mathfrak{D}([H])$ is clear. The converse inclusion is shown as follows. That $\beta \in [H]$ if and only if β is of the form (4.1) for some unitary $v \in \mathfrak{M}$ and some partition $\{e(t) : t \in H\}$ of the identity in $\mathcal{C}_\mathfrak{M}$ such that $\{\alpha_t^{-1}(e(t)) : t \in H\}$ is also a partition of the identity. Setting

$$u = \sum_{t \in H} (1 \otimes \rho(t))\alpha(\alpha_t^{-1}(e(t))v) ,$$

we have $Ad_u \circ \alpha = \alpha \circ \beta$. If $u' \in N(\alpha(\mathfrak{M}))$ and $Ad_{u'} \circ \alpha = \alpha \circ \beta$, then $u^{-1}u' \in \alpha(\mathfrak{M}')$. Since α is free, $u^{-1}u' \in \alpha(\mathcal{C}_\mathfrak{M})$ and so $u' = u\alpha(z)$ for some unitary $z \in \mathcal{C}_\mathfrak{M}$. Hence $u' \in \mathfrak{M} \times_\alpha H$, which means $\mathfrak{D}([H]) \subset \mathfrak{M} \times_\alpha H$. Thus the Dye correspondence gives rise to a finer classification of intermediate von Neumann subalgebras in $\mathcal{R}(\mathfrak{M} \times_\alpha G, \alpha(\mathfrak{M}))$ than that given in Theorem 2.3.

2. Aubert correspondence

Definition 4.4. Let α be an action of G on $\mathfrak{M}$. For each von Neumann subalgebra $\mathcal{P}$ of $\mathfrak{M}$, a partition $\{e_i : i \in I\}$ of the identity in $\mathcal{P}$ is said to be finitely wandering, if

(i) $\{\alpha_t(e_i) : t \in G\}$ is a partition of the identity for each $i \in I$;
(ii) $\alpha_r = \iota$ on $\mathcal{P}$ whenever $\alpha_r(e_i) = e_i$ for some $i \in I$; and
(iii) the set of $r \in G$ with $\alpha_r = \iota$ on $\mathcal{P}$ is finite.

Theorem 4.5. If α is dual, there exists a bijective correspondence of (3.2) between the set of finite subgroups H of G and the set of von Neumann subalgebras $\mathcal{P} \in \mathcal{R}(\mathfrak{M},\mathfrak{M}^\alpha)$ with a finitely wandering partition.

Proof. Since α is dual, we may assume that $\{\mathfrak{M},\alpha\} = \{\mathfrak{N} \times_\delta G, \hat{\delta}\}$ for some covariant system $\{\mathfrak{N},\delta\}$. The correspondence $H(\mathcal{P}(H)) = H$ is a consequence of Theorem 1.2.

Next we shall show $\mathcal{P}(H(\mathcal{P})) = \mathcal{P}$. Since $\mathcal{P} \subset \mathcal{P}(H(\mathcal{P}))$ is clear, it suffices to show the reverse inclusion. Let $\mathcal{E}_\alpha$ be the $\mathfrak{M}^\alpha$-valued weight on $\mathfrak{M}$ defined by (III.1.2):

$$\mathcal{E}_\alpha(x) = \sum_{t \in G} \Delta(t)\alpha_t(x) , \quad x \in \mathfrak{p}_\alpha^+ .$$

Let $\{e_i : i \in I\}$ be a finitely wandering partition. If $t \notin H(\mathcal{P})$, then $\alpha_t(e_i)e_i = 0$ for all $i \in I$ by (ii). Therefore, for any $x \in \mathfrak{M}^{\alpha^H} \cap \mathfrak{p}_\alpha^+$ and $\omega \in \mathfrak{M}_*^+$,

$$\langle \mathcal{E}_\alpha(xe_i)e_i, \omega\rangle = \langle \sum_{t\in G} \Delta(t)\alpha_t(xe_i)e_i, \omega\rangle$$

$$= \langle \sum_{t\in H} \Delta(t)\alpha_t(xe_i)e_i, \omega\rangle = \langle x \sum_{t\in H} \Delta(t)\alpha_t(e_i)e_i, \omega\rangle .$$

Since $\vee \alpha_t(e_i) = 1$ by (i) and H is finite by (iii),

$$\sum_{t\in H} \Delta(t)\alpha_t(e_i)e_i = \mu_i e_i$$

for some $\mu_i > 0$. Therefore

$$xe_i = \mu_i^{-1} \mathcal{E}_\alpha(xe_i)e_i \in \mathcal{P}$$

and hence $x = \sum xe_i \in \mathcal{P}$. Consequently, $\mathfrak{M}^{\alpha^H} \cap \mathfrak{p}_\alpha \subset \mathcal{P}$. Thus it remains to show that $\mathfrak{M}^{\alpha^H} \cap \mathfrak{p}_\alpha$ is σ-weakly dense in $\mathfrak{M}^{\alpha^H}$. As H is finite and $\mathcal{E}_\alpha(\alpha_t(x)) = \Delta(t)^{-1}\mathcal{E}_\alpha(x)$, each element $\sum_{t\in H} \alpha_t(x)$ with $x \in \mathfrak{p}_\alpha$ belongs to $\mathfrak{M}^{\alpha^H} \cap \mathfrak{p}_\alpha$. Since α is dual, $\mathfrak{p}_\alpha$ is σ-weakly dense in $\mathfrak{M}$. Since H is finite, the set of all $\sum_{t\in H} \alpha_t(x)$ with $x \in \mathfrak{p}_\alpha$ is σ-weakly dense in $\mathfrak{M}^{\alpha^H} = \{\sum_{t\in H}\alpha_t(y) : y \in \mathfrak{M}\}$. Thus $\mathfrak{M}^{\alpha^H} \cap \mathfrak{p}_\alpha$ is σ-weakly dense in $\mathfrak{M}^{\alpha^H}$. Q.E.D.

NOTES

The concept of full group is defined for an abelian von Neumann algebra by Dye [21] and generalized to a general von Neumann algebra by Haga and Takeda [37] as in Definition 4.1 and by Connes [12], independently. Theorems 4.2 and 4.3 are obtained in [21] for abelian von Neumann algebras and in [37] for general von Neumann algebras. Theorem 4.5 is obtained by Aubert [5] for a finite group G.

APPENDIX

To a unitary representation $\{u,\mathfrak{H}\}$ of G on a Hilbert space $\mathfrak{H}$ there corresponds bijectively a non-degenerate $*$ representation π of $L^1(G)$ on $\mathfrak{H}$ in such a way that

$$\pi(f) = \int f(t)u(t)dt \ , \quad f \in L^1(G) \ .$$

If G is abelian, then $L^1(\hat{G}) = \mathfrak{F}A(G)\mathfrak{F}^{-1}$. Therefore, a unitary representation of the dual group $\hat{G}$ corresponds to a non-degenerate $*$ representation of the Fourier algebra $A(G)$. The aim of this appendix is to discuss such a representation for general locally compact groups.

<u>Theorem</u> A.1. (a) There exists a bijection between the set of all non-degenerate $*$ representations $\{\pi,\mathfrak{H}\}$ of $L^1(G)$ and the set of all unitaries $u \in \mathfrak{L}(\mathfrak{H}) \bar{\otimes} L^\infty(G)$ with the associativity condition:

$$\text{(A.1)} \qquad (u \otimes 1)(\iota \otimes \sigma)(u \otimes 1) = (\iota \otimes \alpha_G)(u) \ ,$$

which is determined by the relation:

$$\text{(A.2)} \qquad \langle \pi(f),\omega\rangle = \langle u,\omega \otimes f\rangle \ , \quad f \in L^1(G) \ , \quad \omega \in \mathfrak{L}(\mathfrak{H})_* \ .$$

(b) There exists a bijection between the set of all non-degenerate $*$ representations $\{\pi,\mathfrak{K}\}$ of $A(G)$ and the set of all unitaries $w \in \mathfrak{L}(\mathfrak{K}) \bar{\otimes} \mathfrak{R}(G)$ with the associativity condition

$$\text{(A.3)} \qquad (w \otimes 1)(\iota \otimes \sigma)(w \otimes 1) = (\iota \otimes \delta_G)(w)$$

by the relation

$$\text{(A.4)} \qquad \langle \pi(\varphi),\omega\rangle = \langle w,\omega \otimes \varphi\rangle \ , \quad \varphi \in A(G) \ , \quad \omega \in \mathfrak{L}(\mathfrak{K})_* \ .$$

<u>Proof</u>. (a) That a unitary $u \in \mathfrak{L}(\mathfrak{H}) \bar{\otimes} L^\infty(G)$ satisfies (A.1) is equivalent to the existence of a unitary representation $\{U,\mathfrak{H}\}$ with $(u\xi)(t) = U(t)\xi(t)$ for $\xi \in \mathfrak{H} \otimes L^2(G)$. Thus (a) is a known result stated in the above.

(b) The correspondence of w to π: For each $\varphi \in A(G)$ the map: $(\xi,\eta) \in \mathfrak{K} \times \mathfrak{K} \to \langle w,\omega_{\xi,\eta} \otimes \varphi\rangle$ turns out to be a bounded Hermitian form. There exists then a bounded operator $\pi(\varphi)$ in $\mathfrak{L}(\mathfrak{K})$ such that

$$\langle \pi(\varphi),\omega_{\xi,\eta}\rangle = \langle w,\omega_{\xi,\eta} \otimes \varphi\rangle \ , \quad \varphi \in A(G) \ .$$

Clearly, π is a linear map of $A(G)$ into $\mathfrak{L}(\mathfrak{K})$. According to (A.3), we have, for $\varphi,\psi \in A(G)$ and $\omega \in \mathfrak{L}(\mathfrak{K})_*$,

$$\begin{aligned}\langle \pi(\varphi\psi),\omega\rangle &= \langle w,\omega\otimes\varphi\psi\rangle = \langle(\iota\otimes\delta_G)(w),\omega\otimes\varphi\otimes\psi\rangle \\ &= \langle (w\otimes 1)(\iota\otimes\sigma)(w\otimes 1),\omega\otimes\varphi\otimes\psi\rangle \\ &= \langle w\otimes 1,((\iota\otimes\sigma)(w\otimes 1))(\omega\otimes\varphi\otimes\psi)\rangle \\ &= \langle \pi(\varphi)\otimes 1, w(\omega\otimes\psi)\rangle \\ &= \langle w,\omega\pi(\varphi)\otimes\psi\rangle = \langle \pi(\psi),\omega\pi(\varphi)\rangle \\ &= \langle \pi(\varphi)\pi(\psi),\omega\rangle \ .\end{aligned} \tag{A.5}$$

Therefore, π is multiplicative and $\pi(A(G))$ is abelian.

Next we shall show that π is a * representation, namely, $\pi(\bar{\varphi}) = \pi(\varphi)^*$ for $\varphi \in A(G)$. Since

$$\langle (x\otimes 1)w,\omega\otimes\varphi\rangle = \langle \pi(\varphi),\omega x\rangle = \langle x\pi(\varphi),\omega\rangle$$

and

$$\langle w(x\otimes 1),\omega\otimes\varphi\rangle = \langle \pi(\varphi)x,\omega\rangle \quad ,$$

it follows that, $[x\otimes 1,w] = 0$ is equivalent to $x \in \pi(A(G))'$. Since w is unitary, $\pi(A(G))'$ is closed under the adjoint operation. Therefore, $\mathfrak{C} = \pi(A(G))''$ turns out to be an abelian von Neumann algebra. Here we denote the dual map of π by π^*:

$$\langle \pi(\varphi),\omega\rangle = \langle \varphi,\pi^*(\omega)\rangle \ , \quad \omega \in \mathfrak{C} \ .$$

Since π is multiplicative, if ω is a character of $\mathfrak{C}$, then $\pi^*\omega$ is also a character of $A(G)$. Since these characters are self-adjoint, we have

$$\langle \pi(\varphi)^*,\omega\rangle = \overline{\langle \pi(\varphi),\omega\rangle} = \overline{\langle \varphi,\pi^*\omega\rangle} = \langle \bar{\varphi},\pi^*\omega\rangle = \langle \pi(\bar{\varphi}),\omega\rangle \ .$$

This mean that $\pi(\varphi)^* = \pi(\bar{\varphi})$.

Finally we shall show that the * representation π is non-degenerate. Suppose that $\pi(A(G))\xi = 0$ for $\xi \in \mathfrak{K}$. For any $\eta \in \mathfrak{K}$, $f, g \in L^2(G)$, we have

$$(w(\xi\otimes f)|\eta\otimes g) = \langle \pi(\omega_{f,g}),\omega_{\xi,\eta}\rangle = 0 \quad .$$

Since w is unitary, $\xi\otimes f = 0$ for any $f \in L^2(G)$. Therefore, $\xi = 0$, i.e. π is is non-degenerate.

The correspondence of π to w: Let $\{\pi,\mathfrak{K}\}$ be a non-degenerate * representation of $A(G)$. Since $\mathfrak{C} = \pi(A(G))''$ is abelian, it follows that $\mathfrak{C} \hat{\otimes}_\alpha \mathfrak{R}(G) = \mathfrak{C} \hat{\otimes}_\lambda \mathfrak{R}(G)$, [62]. Therefore

$$\pi \in \mathcal{L}(A(G),\mathfrak{A}) = (\mathfrak{A}_* \hat{\otimes}_\gamma A(G))^* = \mathfrak{A} \hat{\otimes}_\lambda \mathfrak{R}(G) = \mathfrak{A} \bar{\otimes} \mathfrak{R}(G) .$$

Here we denote the element in $\mathfrak{A} \bar{\otimes} \mathfrak{R}(G)$ corresponding to π by w. Then we have (A.4):

$$\langle \pi(\varphi),\omega \rangle = \langle w,\omega \otimes \varphi \rangle , \quad \varphi \in A(G) , \quad \omega \in \mathcal{L}(\mathfrak{K})_* .$$

Making use of the multiplicativity of π and the similar computation as (A.5), we get the associativity condition (A.3). It remains to show that w is unitary.

Now, the $*$ representation $\{\pi,\mathfrak{K}\}$ of $A(G)$ is extended uniquely to a $*$ representation of $C_\infty(G) = C^*(A(G))$, which is denoted by the same symbol $\{\pi,\mathfrak{K}\}$. Since, for any $\xi,\eta \in \mathfrak{K}$ and $f,g \in \mathcal{K}(G)$,

$$\begin{aligned}(w(\xi \otimes f) \mid \eta \otimes g) &= \langle w,\omega_{\xi,\eta} \otimes \omega_{f,g}\rangle \\ &= \langle \pi(g^\# * f),\omega_{\xi,\eta}\rangle = \int \langle \pi(\lambda_t^{-1}f),\omega_{\xi,\eta}\rangle \overline{g(t)}dt ,\end{aligned}$$

we have

$$(w(\xi \otimes f))(t) = \pi(\lambda_t^{-1}f)\xi , \text{ locally a.e. in } t . \tag{A.6}$$

Therefore

$$\begin{aligned}(w(\xi \otimes f) \mid w(\eta \otimes g)) &= \int (\pi(\lambda_t^{-1}f)\xi \mid \pi(\lambda_t^{-1}g)\eta)dt \\ &= \int \langle \pi(\lambda_t^{-1}(\bar{g}f)),\omega_{\xi,\eta}\rangle dt .\end{aligned} \tag{A.7}$$

For each $\omega \in \mathcal{L}(\mathfrak{K})_*$, the map: $f \in C_\infty(G) \to \langle \pi(f),\omega\rangle$ is a bounded linear functional, that is $\pi^*\omega$ is a Radon measure on G: $\langle f,\pi^*\omega\rangle = \langle \pi(f),\omega\rangle$. Therefore

$$\begin{aligned}\int \langle \pi(\lambda_t^{-1}(\bar{g}f)),\omega_{\xi,\eta}\rangle dt &= \iint (\bar{g}f)(ts)d\pi^*\omega_{\xi,\eta}(s)dt \\ &= (f \mid g) \int d\pi^*\omega_{\xi,\eta}(s) ,\end{aligned} \tag{A.8}$$

where the last equality is obtained by Fubini Theorem, for $\bar{g}f \in \mathcal{K}(G)$. On the other hand, π is non-degenerate. So the weak closure of the set of $\pi(f)$ with $f \in C_\infty(G)$, $0 \leq f \leq 1$ contains the identity. Since π is a $*$ representation, we can choose f so that $f \nearrow 1$ implies $\pi(f) \nearrow 1$. Therefore, the first equality of (A.8) gives us

$$(\xi \mid \eta) = \int d\pi^*\omega_{\xi,\eta}(s) . \tag{A.9}$$

Combining (A.7), (A.8) and (A.9), we find that w is an isometry. As for w^* we have the similar equality as (A.6):

$$(w^*(\eta \otimes g))(t) = \pi((\lambda_t^{-1} g)^\vee)\eta$$

in place of it. Repeating the same argument as above, we find that w^* is an isometry. Q.E.D.

Example 1. If $\{w, \mathfrak{K}\} = \{W_G, L^2(G)\}$, then w satisfies (A.3). The corresponding non-degenerate $*$ representation $\{\pi, L^2(G)\}$ of $A(G)$ is given as the multiplication operators on $L^2(G)$:

$$\pi(\varphi)\xi = \varphi\xi \ , \quad \xi \in L^2(G) \ .$$

Indeed, $\langle \pi(\omega_{f,g}), \omega_{\xi,\eta} \rangle = (W_G(\xi \otimes f) \mid \eta \otimes g) = ((g^\# * f)\xi \mid \eta)$.

Example 2. If G is discrete, there exists a partition $\{e(t) : t \in G\}$ of the identity such that $w = \sum_{t \in G} e(t) \otimes \rho(t)$. The corresponding representation π is given by $\pi(\varphi) = \sum_{t \in G} \varphi(t) e(t)$.

For any unitary representations u^j of G on $\mathfrak{H}_j$, or unitaries u^j in $\mathfrak{L}(\mathfrak{H}_j) \mathbin{\bar{\otimes}} L^\infty(G)$ satisfying (A.1), we consider a representation:

$$t \to u^1(t) \otimes u^2(t)$$

of G on $\mathfrak{H}_1 \otimes \mathfrak{H}_2$, which is also a unitary in $\mathfrak{L}(\mathfrak{H}_1 \otimes \mathfrak{H}_2) \otimes L^\infty(G)$ denoted by $u^1 * u^2$. Then

(A.10)
$$\begin{aligned} u^1 * u^2 &= (1 \otimes u^2)(\iota \otimes \sigma)(u^1 \otimes 1) \\ &= ((\iota \otimes \sigma)(u^1 \otimes 1))(1 \otimes u^2) \end{aligned}$$

on $\mathfrak{H}_1 \otimes \mathfrak{H}_2 \otimes L^2(G)$. In § 3 of Chapter I we use the notation $u^1 \otimes u^2$ for the representation $t \to u^1(t) \otimes u^2(t)$ instead of $u^1 * u^2$.

Definition A.2. Let π_j be non-degenerate $*$ representations of $A(G)$ on $\mathfrak{K}_j$ for $j = 1,2$. The product $\pi_1 * \pi_2$ of π_1 and π_2 is defined by

(A.11)
$$\langle \pi_1 * \pi_2(\varphi), \omega_1 \otimes \omega_2 \rangle = \langle \varphi, \pi_1^* \omega_1 * \pi_2^* \omega_2 \rangle$$

for $\varphi \in A(G)$ and $\omega_j \in \mathfrak{L}(\mathfrak{K}_j)_*$.

Theorem A.3. If $\pi_j (j = 1,2)$ are non-degenerate $*$ representations of $A(G)$ on $\mathfrak{K}_j$, then $\pi_1 * \pi_2$ is a non-degenerate $*$ representation of $A(G)$ on $\mathfrak{K}_1 \otimes \mathfrak{K}_2$ and satisfies that

(A.12)
$$w_{\pi_1 * \pi_2} = ((\iota \otimes \sigma)(w_{\pi_1} \otimes 1))(1 \otimes w_{\pi_2}) \ ,$$

(A.13)
$$(\pi_1 * \pi_2)(g^\# * f) = \int \pi_1(\rho_t^{-1} g^\#) \otimes \pi_2(\lambda_t^{-1} f) dt \ , \quad f, g \in K(G) \ ,$$

where w_π is the unitary corresponding to π by (A.4).

<u>Proof</u>. Put $w_1 = w_{\pi_1}$, $w_2 = w_{\pi_2}$ and $w = ((\iota \otimes \sigma)(w_1 \otimes 1))(1 \otimes w_2)$. Then w is a unitary in $\mathcal{L}(\mathfrak{K}_1 \otimes \mathfrak{K}_2) \bar{\otimes} L^\infty(G)$. Since

$$(\iota \otimes \sigma \otimes \iota) \circ (\iota \otimes \iota \otimes \sigma) \circ (\iota \otimes \delta_G \otimes \iota) = (\iota \otimes \iota \otimes \delta_G) \circ (\iota \otimes \sigma) ,$$

it follows from (A.3) that

$$\begin{aligned}&(\iota \otimes \sigma \otimes \iota)((w_1 \otimes 1 \otimes 1)(\iota \otimes \sigma \otimes \iota) \circ (\iota \otimes \iota \otimes \sigma) \circ (\iota \otimes \sigma \otimes \iota)(w_1 \otimes 1 \otimes 1))\\ &= (\iota \otimes \sigma \otimes \iota) \circ (\iota \otimes \iota \otimes \sigma)((w_1 \otimes 1 \otimes 1)(\iota \otimes \sigma \otimes \iota)(w_1 \otimes 1 \otimes 1))\\ &= (\iota \otimes \iota \otimes \delta_G) \circ (\iota \otimes \sigma)(w_1 \otimes 1) .\end{aligned}$$

Again, by (A.3), we have

$$(1 \otimes w_2 \otimes 1)(\iota \otimes \iota \otimes \sigma)(1 \otimes w_2 \otimes 1) = (\iota \otimes \iota \otimes \delta_G)(1 \otimes w_2) .$$

Therefore w satisfies the associativity condition:

$$\begin{aligned}&(w \otimes 1)(\iota \otimes \iota \otimes \sigma)(w \otimes 1)\\ &= ((\iota \otimes \sigma \otimes \iota)(w_1 \otimes 1 \otimes 1))(1 \otimes w_2 \otimes 1)\\ &\qquad \cdot ((\iota \otimes \iota \otimes \sigma) \circ (\iota \otimes \sigma \otimes \iota)(w_1 \otimes 1 \otimes 1))(\iota \otimes \iota \otimes \sigma)(1 \otimes w_2 \otimes 1)\\ &= ((\iota \otimes \iota \otimes \delta_G) \circ (\iota \otimes \sigma)(w_1 \otimes 1))(\iota \otimes \iota \otimes \delta_G)(1 \otimes w_2)\\ &= (\iota \otimes \iota \otimes \delta_G)(w) .\end{aligned}$$

Now, for any $\varphi \in A(G)$ and $\omega_j \in \mathcal{L}(\mathfrak{K}_j)_*$, we have

$$\begin{aligned}\langle \varphi, \pi_1^*\omega_1 * \pi_2^*\omega_2 \rangle &= \iint \varphi(st)d\pi_1^*\omega_1(s)d\pi_2^*\omega_2(t)\\ &= \int \langle w_1, \omega_1 \otimes \rho_t\varphi \rangle d\pi_2^*\omega_2(t)\\ &= \int \langle 1 \otimes \rho(t), (\omega_1 \otimes \varphi)w_1 \rangle d\pi_2^*\omega_2(t) ,\end{aligned}$$

where the third equality follows from

$$(\rho_t\varphi)(s) = \langle \rho(st), \varphi \rangle = \langle \rho(s), \rho(t)\varphi \rangle .$$

Since $(\omega_1 \otimes \varphi)w_1$ is of the form $\sum_{j\geq 3} \omega_j \otimes \varphi_j$ for some $\omega_j \in \mathcal{L}(\mathfrak{K}_1)_*$ and $\varphi_j \in A(G)$, it follows that

$$\langle \varphi, \pi_1^*\omega_1 * \pi_2^*\omega_2 \rangle = \int \sum_{j\geq 3} \langle 1 \otimes \rho(t), \omega_j \otimes \varphi_j \rangle d\pi_2^*\omega_2(t)$$

$$= \sum_{j\geq 3} \langle 1, \omega_j \rangle \int \langle \rho(t), \varphi_j \rangle d\pi_2^*\omega_2(t)$$

$$= \sum_{j\geq 3} \langle 1, \omega_j \rangle \langle w_2, \omega_2 \otimes \varphi_j \rangle$$

$$= \langle 1 \otimes w_2, \sum_{j\geq 3} \omega_j \otimes \omega_2 \otimes \varphi_j \rangle$$

$$= \langle ((\iota \otimes \sigma)(w_1 \otimes 1))(1 \otimes w_2), \omega_1 \otimes \omega_2 \otimes \varphi \rangle$$

$$= \langle w, \omega_1 \otimes \omega_2 \otimes \varphi \rangle \quad .$$

Therefore, $\pi_1 * \pi_2$ is a non-degenerate * representation of $A(G)$ with (A.12).

It remains to show (A.13). For any $f, g \in K(G)$ and $\omega_j \in \mathcal{L}(\mathfrak{K}_j)_*$ we have

$$\langle (\pi_1 * \pi_2)(g^\# * f), \omega_1 \otimes \omega_2 \rangle = \iint (g^\# * f)(rs) d\pi_1^*\omega_1(r) d\pi_2^*\omega_2(s)$$

$$= \iiint g^\#(rt^{-1}) f(ts) dt \; d\pi_1^*\omega_1(r) d\pi_2^*\omega_2(s) \quad .$$

Since $f, g^\# \in K(G)$, we use Fubini theorem and

$$\langle (\pi_1 * \pi_2)(g^\# * f), \omega_1 \otimes \omega_2 \rangle$$

$$= \int \langle \pi_1(\rho_t^{-1} g^\#), \omega_1 \rangle \langle \pi_2(\lambda_t^{-1} f), \omega_2 \rangle dt \quad ,$$

which implies (A.13). Q.E.D.

In the above theorem if $s_j (j = 1,2)$ are one dimensional non-degenerate * representations of $A(G)$ on $\mathfrak{K}_j$, then they are characters of $A(G)$ and

$$(s_1 * s_2)(g^\# * f) = \int s_1(\rho_t^{-1} g^\#) \otimes s_2(\lambda_t^{-1} f) dt$$

$$= \int g^\#(s_1 t^{-1}) f(ts_2) dt = (g^\# * f)(s_1 s_2) \quad .$$

<u>Theorem</u> A.4. If δ is a co-action of G on $\mathfrak{h}$ implemented by a unitary $w \in \mathfrak{h} \bar{\otimes} \mathcal{R}(G)$ with (A.3) and π is the non-degenerate * representation of $A(G)$ corresponding to w, then

$$\delta_{\varphi}(y) = \int \pi(\lambda_t^{-1}g)^* y \, \pi(\lambda_t^{-1}f)dt \ , \quad (\varphi = \omega_{f,g}) \ .$$

<u>Proof</u>. It is clear by (A.6). Q.E.D.

REFERENCES

[1] Araki, H.; Some properties of modular conjugation operator of von Neumann algebras and a non commutative Radon-Nikodym theorem with a chain rule, Pacific J. Math, 50(1974), 309-354.

[2] Araki, H.; Structure of some von Neumann algebras with isolated discrete modular spectrum, Publ. RIMS, Kyoto Univ., 9(1973), 1-14.

[3] Araki, H., Kastler, D., Takesaki, M. and Haag, R.; Extension of KMS states and chemical potential, Commun. Math. Phys., 53(1977), 97-134.

[4] Arveson, W.B.; On groups of automorphisms of operator algebras, J. Functional Analysis, 15(1974), 217-243.

[5] Aubert, P.L.; Théorie de Galois pour une W*-algébre, Comment. Math. Helvetici, 39(51)(1976), 411-433.

[6] Choda, H.; A Galois correspondence in a von Neumann algebra, Preprint of Osaka Kyoiku Univ., 1977.

[7] Choda, H.; Correspondences between von Neumann algebras and discrete automorphism groups, The 2nd Japan-US Seminar on C*-algebras and its Applications to Physics, 1977.

[8] Choda, M.; Shift automorphism groups of von Neumann algebras, Proc. Japan Acad., 50(1974), 470-475.

[9] Choda, M.; Normal expectations and crossed products of von Neumann algebras, Proc. Japan Acad., 50(1974), 738-742.

[10] Choda, M.; The fixed algebra of a von Neumann algebra under an automorphism group, Tôhoku Math. J., 28(1976), 227-234.

[11] Choda, M.; Correspondence between subgroups and subalgebras in the crossed product of a von Neumann algebra, Math. Japonicae, 21(1976), 51-59.

[12] Connes, A.; Une classification des facteurs de type III, Ann. Sci. École Norm. Sup., 6(1973), 133-252.

[13] Connes, A. and Størmer, E.; Homogeneity of the state space of factors of type III, J. Functional Analysis, 28(1978), 187-196.

[14] Connes, A. and Takesaki, M.; The flow of weights on factors of type III, Tôhoku Math. J., 29(1977), 473-575.

[15] Digerness, T.; Poids duals sur un produit croisé, C.R. Acad. Sc. Paris, 278 (1974), A937-A940.

[16] Digerness, T.; Dual weight and the commutant theorem for crossed products of W*-algebras, Preprint of UCLA, 1974.

[17] Doplicher, S., Haag, R. and Roberts, J.E.; Fields, observables and gauge transformations I, Commun. Math. Phys., 13(1969), 1-13; II, ibid., 15(1969), 173-201.

[18] Doplicher, S., Haag, R. and Roberts, J.E.; Local observables and particle statistics I, Commun. Math. Phys., 23(1971), 199-230; II, ibid., 35(1974), 49-85.

[19] Doplicher, S., Haag, R. and Roberts, J.E.; Local observables and particle statistics and non-abelian gauge groups, Commun. Math Phys., 28(1972), 331-348.

[20] Doplicher, S., Kastler, D. and Robinson, D.; Covariance algebras in field theory and statistical Mechanics, Commun. Math. Phys., 3(1966), 1-28.

[21] Dye, H.; On groups of measure preserving transformations, I, Amer. J. Math., 81(1959), 119-159; II, ibid, 85(1963), 551-576.

[22] Enock, M.; Produit croise d'une algèbre de von Neumann par algèbre de Kac, J. Functional Analysis, 26(1977), 16-47.

[23] Enock, M. and Schwartz, J.M.; Une dualité dans les algèbres de von Neumann, Bull. Soc. Math. France, Suppl. Memoire, 44(1975), 1-144.

[24] Enock, M. and Schwartz, J.M.; Une nouvelle construction du poids dual sur le produit croisé d'une algèbre de von Neumann par un groupe localement compact, C. R. Acad. Sc. Paris, 282(1976), A415-A418.

[25] Enock, M. and Schwartz, J.M.; Algèbre de Kac et produits croisés, C. R. Acad. Sc. Paris, 283(1976), A321-A323.

[26] Enock, M. and Schwartz, J.M.; Produit croisé d'une algèbre de von Neumann par algèbre de Kac, II, Publ. RIMS, Kyoto Univ., to appear.

[27] Ernest, J.; Hopf-von Neumann algebras, Functional Analysis Conf. Irvine 1966, Academic Press 1967, 195-215.

[28] Eymard, P.; L'algebre de Fourier d'un groupe localement compact, Bull. Soc. Math. France, 92(1964), 181-236.

[29] Fell, J.M.G.; An extension of Mackey's method to Banach *-algebraic bundles, Memoir of the Amer. Math. Soc., 90(1969), 1-168.

[30] Fell, J.M.G.; Induced representations and Banach *-algebraic bundles, with an appendix due to A. Douady and L. Dal Soglio-Herault, Lecture Notes in Math., 582(1977), 1-349, Springer-Verlag.

[31] Golodets, V. Ya.; Crossed products of von Neumann algebras, Russian Math. Surveys, 26(1971), 1-50.

[32] Haagerup, U.; The standard form of von Neumann algebras, Math. Scand., 37(1975), 271-283.

[33] Haagerup, U.; Operator valued weights in von Neumann algebras, Preprint of Odense Univ. 1975.

[34] Haagerup, U.; On the dual weights for crossed products of von Neumann algebras, I, II, Preprints of Odense Univ. 1976.

[35] Haga, Y.; On subalgebras of a von Neumann algebra, Tôhoku Math. J., 25(1973), 291-305.

[36] Haga, Y.; Crossed products of von Neumann algebras by compact groups, Tôhoku Math. J., 28(1976), 511-522.

[37] Haga, Y. and Takeda, Z.; Correspondence between subgroups and subalgebras in a cross product von Neumann algebra, Tôhoku Math. J., 24(1972), 167-190.

[38] Ikunishi, A. and Nakagami, Y.; On invariant $G(\sigma)$ and $\Gamma(\sigma)$ for an automorphism group of a von Neumann algebra, Publ. RIMS, Kyoto Univ., 12(1976) 1-30.

[39] Kallman, R.; A generalization of free action, Duke Math. J., 36(1969), 781-789.

[40] Kirchberg, E.; Darstellungen coinvolutiver Hopf-W*-Algebren und ihre Anwendung in der nicht-abelschen Dualitäts theorie Lokalkompakter Gruppen, Berlin, 1977, pp. 354.

[41] Kishimoto, A.; Remarks on compact automorphism groups of a certain von Neumann algebra, Publ. RIMS, Kyoto Univ., 13(1977), 573-581.

[42] Landstad, M.B.; Duality theory for covariant systems, Trans. Amer. Math. Soc., to appear.

[43] Landstad, M.B.; Duality for dual covariance algebras, Commun. Math. Phys., 52 (1977), 191-202.

[44] Mackey, G.W.; A theorem of Stone and von Neumann, Duke Math. J., 16(1949), 313-326.

[45] Nakagami, Y.; Duality for crossed products of von Neumann algebras by locally compact groups, Bull. Amer. Math. Soc., 81(1975), 1106-1108.

[46] Nakagami, Y.; Dual action on a von Neumann algebra and Takesaki's duality for a locally compact group, Publ, RIMS, Kyoto Univ., 12(1977), 727-775.

[47] Nakagami, Y.; Essential spectrum $\Gamma(\beta)$ of a dual action on a von Neumann algebra, Pacific J. Math., 70(1977), 437-479.

[48] Nakagami, Y. and Oka, Y.; On Connes spectrum Γ of a tensor product of actions on von Neumann algebras, Yokohama Math. J., 26(1978), to appear.

[49] Nakagami, Y. and Sutherland, C.; Takesaki's duality for regular extensions of von Neumann algebras, The 2nd Japan-US Seminar on C*-algebras and Applications to Physics, 1977.

[50] Nakagami, Y. and Sutherland, C.; Takesaki's duality for regular extensions of von Neumann algebras, Pacific J. Math., to appear.

[51] Nakamura, M. and Takeda, Z.; On some elementary properties of the crossed product of von Neumann algebras, Proc. Japan Acad., 34(1958), 489-494.

[52] Nakamura, M. and Takeda, Z.; A Galois theory for finite factors, Proc. Japan Acad., 36(1960), 258-260.

[53] Paschke, W.L.; Integrable group actions on von Neumann algebras, Math. Scand., 40(1977), 234-248.

[54] Paschke, W.L.; Relative commutant of a von Neumann algebra in its crossed product by a group action, Preprint, 1977.

[55] Roberts, J.E.; Cross products of a von Neumann algebras by group duals, Inst. Nazional di Alta Mat. Symposia Math., 20(1976), Academic Press, 1977.

[56] Saito, K.; On a duality for locally compact groups, Tôhoku Math. J., 20(1968), 355-367.

[57] Sauvageot, J.L.; Sur le type du produit croisé d'une algèbre de von Neumann par un groupe localement compact d'automorphismes, C.R. Acad. Sc. Paris, 278(1974), A941-A944; Bull. Soc. Math. France, 105(1977), 349-368.

[58] Schwartz, J.M.; Sur la structure des algebres de Kac, Preprint, 2nd edition, 1977.

[59] Strătilă, D., Voiculescu, D. and Zsidó, L.; Sur les produits croisés, C. R. Acad. Sc. Paris, 280(1975), A555-A558.

[60] Strătilă, D., Voiculescu, D. and Zsidó, L.; On crossed products, I, Rev. Roumaine Math., 21(1976), 1411-1449; II, ibid., 22(1977), 83-117.

[61] Suzuki, N.; Crossed products of rings of operators, Tôhoku Math. J., 11(1959), 113-124.

[62] Takesaki, M.; A note on the cross-norm of the direct product of operator algebras, Kōdai Math. Sem. Rep., 10(1958), 137-140.

[63] Takesaki, M.; A generalized commutation relation for the regular representation, Bull. Soc. Math. France, 97(1969), 289-297.

[64] Takesaki, M.; A characterization of group algebras as a converse of Tanaka-Steinspring-Tatsuuma duality theorem, Amer.J.Math., 91(1969), 529-564.

[65] Takesaki, M.; Covariant representations of C*-algebras and their locally compact automorphism groups, Acta Math., 131(1971), 249-310.

[66] Takesaki, M.; Duality and von Neumann algebras, Lecture on Operator Algebras (Edited by K.H. Hofmann), Lecture notes in Math., 247(1972), 666-779, Springer-Verlag.

[67] Takesaki, M.; Dualité dans les produits croisés d'algèbres de von Neumann, C. R. Acad. Sc. Paris, 276(1973), A41-A43.

[68] Takesaki, M.; Duality in crossed products and von Neumann algebras of type III, Bull. Amer. Math. Soc., 79(1973), 1004-1005.

[69] Takesaki, M.; Duality in crossed products and the structure of von Neumann algebras of type III, Acta Math., 131(1973), 249-310.

[70] Takesaki, M.; Relative commutant theorem in crossed products and the exact sequence for the automorphism group of a factor of type III, Inst. Nazionale di Alta Mat. Symposia Math., 20(1976), Academic Press, 1977.

[71] Takesaki, M. and Tatsuuma, N.; Duality and subgroups, Ann. of Math., 93(1971), 344-364.

[73] Tatsuuma, N.; A duality theorem for locally compact groups, J. Math. Kyoto Univ., 6(1967), 187-293.

[74] Tatsuuma, N.; An extension of AKTH-theory to locally compact groups, Preprint, 1977.

[75] Tomiyama, J.; Tensor products and projections of norm one in von Neumann algebras, Lecture note at Univ. Copenhagen, Denmark, 1970.

[76] Turumaru, T.; Crossed product of operator algebra, Tôhoku Math. J., 10(1958), 355-365.

[77] Van Heeswijck, L.; Duality in the theory of crossed products, Preprint of Katholicke Univ. Leuven, 1978.

[78] Walter, M.; W*-algebras and non-abelian harmonic analysis, J. Functional Analysis, 11(1972), 17-38.

[79] Zeller-Meier, G.; Produits croisés d'une C*-algèbre par un groupe d'automorphisms, J. Math. pures et appl., 47(1968), 101-239.

[80] Takesaki, M.; Tomita's theory of modular Hilbert algebras and its applications, Lecture Notes in Math., 128(1970), 1-123, Springer-Verlag.

[81] Takesaki, M.; Lecture on operator algebras at UCLA, (1970/71).

Vol. 700: Module Theory, Proceedings, 1977. Edited by C. Faith and S. Wiegand. X, 239 pages. 1979.

Vol. 701: Functional Analysis Methods in Numerical Analysis, Proceedings, 1977. Edited by M. Zuhair Nashed. VII, 333 pages. 1979.

Vol. 702: Yuri N. Bibikov, Local Theory of Nonlinear Analytic Ordinary Differential Equations. IX, 147 pages. 1979.

Vol. 703: Equadiff IV, Proceedings, 1977. Edited by J. Fábera. XIX, 441 pages. 1979.

Vol. 704: Computing Methods in Applied Sciences and Engineering, 1977, I. Proceedings, 1977. Edited by R. Glowinski and J. L. Lions. VI, 391 pages. 1979.

Vol. 705: O. Forster und K. Knorr, Konstruktion verseller Familien kompakter komplexer Räume. VII, 141 Seiten. 1979.

Vol. 706: Probability Measures on Groups, Proceedings, 1978. Edited by H. Heyer. XIII, 348 pages. 1979.

Vol. 707: R. Zielke, Discontinuous Čebyšev Systems. VI, 111 pages. 1979.

Vol. 708: J. P. Jouanolou, Equations de Pfaff algébriques. V, 255 pages. 1979.

Vol. 709: Probability in Banach Spaces II. Proceedings, 1978. Edited by A. Beck. V, 205 pages. 1979.

Vol. 710: Séminaire Bourbaki vol. 1977/78, Exposés 507–524. IV, 328 pages. 1979.

Vol. 711: Asymptotic Analysis. Edited by F. Verhulst. V, 240 pages. 1979.

Vol. 712: Equations Différentielles et Systèmes de Pfaff dans le Champ Complexe. Edité par R. Gérard et J.-P. Ramis. V, 364 pages. 1979.

Vol. 713: Séminaire de Théorie du Potentiel, Paris No. 4. Edité par F. Hirsch et G. Mokobodzki. VII, 281 pages. 1979.

Vol. 714: J. Jacod, Calcul Stochastique et Problèmes de Martingales. X, 539 pages. 1979.

Vol. 715: Inder Bir S. Passi, Group Rings and Their Augmentation Ideals. VI, 137 pages. 1979.

Vol. 716: M. A. Scheunert, The Theory of Lie Superalgebras. X, 271 pages. 1979.

Vol. 717: Grosser, Bidualräume und Vervollständigungen von Banachmoduln. III, 209 pages. 1979.

Vol. 718: J. Ferrante and C. W. Rackoff, The Computational Complexity of Logical Theories. X, 243 pages. 1979.

Vol. 719: Categorial Topology, Proceedings, 1978. Edited by H. Herrlich and G. Preuß. XII, 420 pages. 1979.

Vol. 720: E. Dubinsky, The Structure of Nuclear Fréchet Spaces. V, 187 pages. 1979.

Vol. 721: Séminaire de Probabilités XIII. Proceedings, Strasbourg, 1977/78. Edité par C. Dellacherie, P. A. Meyer et M. Weil. VII, 647 pages. 1979.

Vol. 722: Topology of Low-Dimensional Manifolds. Proceedings, 1977. Edited by R. Fenn. VI, 154 pages. 1979.

Vol. 723: W. Brandal, Commutative Rings whose Finitely Generated Modules Decompose. II, 116 pages. 1979.

Vol. 724: D. Griffeath, Additive and Cancellative Interacting Particle Systems. V, 108 pages. 1979.

Vol. 725: Algèbres d'Opérateurs. Proceedings, 1978. Edité par P. de la Harpe. VII, 309 pages. 1979.

Vol. 726: Y.-C. Wong, Schwartz Spaces, Nuclear Spaces and Tensor Products. VI, 418 pages. 1979.

Vol. 727: Y. Saito, Spectral Representations for Schrödinger Operators With Long-Range Potentials. V, 149 pages. 1979.

Vol. 728: Non-Commutative Harmonic Analysis. Proceedings, 1978. Edited by J. Carmona and M. Vergne. V, 244 pages. 1979.

Vol. 730: Functional Differential Equations and Approximation of Fixed Points. Proceedings, 1978. Edited by H.-O. Peitgen and H.-O. Walther. XV, 503 pages. 1979.

Vol. 731: Y. Nakagami and M. Takesaki, Duality for Crossed Products of von Neumann Algebras. IX, 139 pages. 1979.